Progress in Probability and Statistics

Volume 15

Series Editor

Murray Rosenblatt

Seminar on Stochastic Processes, 1987

E. Çınlar,
K.L. Chung,
R.K. Getoor,
Editors

J. Glover,
Managing Editor

1988

Birkhäuser
Boston · Basel

E. Çınlar
Civil Engineering Department
Princeton University
Princeton, NJ 08544
U.S.A.

K.L. Chung
Department of Mathematics
Stanford University
Stanford, CA 94305
U.S.A.

R.K. Getoor
Department of Mathematics
University of California
La Jolla, CA 92093
U.S.A.

J. Glover (Managing Editor)
Department of Mathematics
University of Florida
Gainesville, FL 32611
U.S.A.

ISSN: 0892-063X

CIP-Titelaufnahme der Deutschen Bibliothek
Seminar on Stochastic Processes:
Seminar on Stochastic Processes . . . — Boston ; Basel :
Birkhäuser.
Bis 6. 1986 mit d. Erscheinungsorten Boston, Basel, Stuttgart
WG: 27;15 DBN 55.092348.9
5259/01 bg
7. 1987 (1988)
(Progress in probability and statistics ; Vol. 15)
ISBN-13: 978-1-4684-0552-1 e-ISBN-13: 978-1-4684-0550-7
DOI: 10.1007/978-1-4684-0550-7
NE: GT
WG: 27 DBN 88.010688.3 87.12.21
5259/02* bg

ISBN-13: 978-1-4684-0552-1

Text prepared by the authors in camera-ready form.

9 8 7 6 5 4 3 2 1

Dedication

Seven has always been an auspicious number in the Orient. As the ancients would explain it, it is the five planets and the sun and the moon. It gives my oriental mind a certain pleasure to note that this seminar, the seventh in the series, brought us together to be with Professors Kai Lai Chung and Gilbert Agnew Hunt for a few days in the seventieth year of their distinguished lives.

In fact, the idea of holding these seminars was hatched here, seven years ago, during a visit by Professor Chung. I remember well his stories of the late forties in probability in Princeton and of the events of those days presaging the roles the two classmates were to play in the now well-known tale of probabilities and potentials.

Their fundamental contributions have inspired us all, pointed the way, and established the standards. This series of seminars rings with the clearest expressions of our acknowledgement of our indebtedness to their pioneering spirits. This volume is dedicated to them as a further token of our appreciation and in affectionate tribute to their leadership.

Princeton, 1987 Erhan Çınlar

FOREWORD

The 1987 Seminar on Stochastic Processes was held at Princeton University, March 26 through March 28, 1987. It was the seventh seminar in a continuing series of meetings which provide opportunities for researchers to discuss current work in stochastic processes in an informal and enjoyable atmosphere. Previous seminars were held at Northwestern University, Evanston; University of Florida, Gainesville: and University of Virginia, Charlottesville.

The success of these seminars has been due to the interest and enthusiasm of probabilists in the United States and abroad. Many of the participants have allowed us to publish the results of their research in this volume. The editors hope that the reader will be able to sense some of the excitement present in the seminar by reading these articles. This year's invited participants included M. Aizenman, B. Atkinson, R. M. Blumenthal, C. Burdzy, D. Burkholder, R. Carmona, K. L. Chung, M. Cranston, C. Dellacherie, J. D. Deuschel, N. Dinculeanu, E. B. Dynkin, P. Fitzsimmons, R. K. Getoor, J. Glover, R. Gundy, P. Hsu, G. A. Hunt, H. Kaspi, F. Knight, G. Lawler, P. March, P. A. Meyer, J. Mitro, J. Neveu, E. Pardoux, M. Pinsky, L. Pitt, A. O. Pittenger, Z. Pop-Stojanovic, P. Protter, M. Rao, T. Salisbury, M. J. Sharpe, S. J. Taylor, E. Toby, S. R. S. Varadhan, R. Williams, M. Weber, and Z. Zhao.

The seminar was made possible through the generous support of the Office of Naval Research (Grant No. N00014-87-G-0138) and Princeton University, and we are grateful for their support. Erhan Cinlar was host during the meeting. The pleasant and stimulating ambience of this meeting (and of several past meetings) was due to his efforts, and we extend our thanks to him.

J. G.

Gainesville, 1987

TABLE OF CONTENTS

HOMOGENEITY FOR TWO-SIDED DISCRETE MARKOV PROCESSES

by

BRUCE W. ATKINSON

1. INTRODUCTION.

We consider here three types of homogeneity for Markov processes indexed by the integers with random times of birth and death, and which take values in a countable state space S. The measures on the path space corresponding to such processes are sigma-finite but not necessarily probability measures. However, via the Radon-Nikodym theorem, no difficulty arises in extending the definitions of conditional expectation and conditional independence.

Before describing the main results of this paper, it will be necessary to make some definitions.

(1.1) DEFINITIONS. Let a, b be distinct elements not in S.

(a) $W = \{$functions $w : Z \to S \cup \{a,b\}$: there exists $n \in Z$ with $w(n) \in S$, $w(m) = a \Rightarrow w(k) = a \ \forall\, k < m$, and $w(m) = b \Rightarrow w(k) = b \ \forall\, k > m\}$.

(b) For $n \in Z$, define $x(n) : W \to S \cup \{a,b\}$ by $x(n)(w) = w(n)$.

(c) For $n \in Z$, define $\theta_n : W \to W$ by $\theta_n(w)(m) = w(m+n)$.

(d) A non-negative measure P on $(W, \sigma(x(n), n \in Z))$ is called a <u>two-sided process</u> if $P(x(n) = i) < \infty \ \forall\, i \in Z$. (Note: All processes in this paper are two-sided and so we will drop the implicit prefix "two-sided" from now on.)

(e) A process P is called <u>Markov</u> if there exists a family $\{p_n : n \in Z\}$ of substochastic matrices on S so that whenever $m, n \in Z$, $m \geq 1$, and $i_0, i_1, \ldots, i_m \in S$, then

$$(1.2) \quad P(x(n+k) = i_k \,,\ 0 \leq k \leq m) = P(x(n) = i_0) \prod_{k=0}^{m-1} p_{n+k}(i_k, i_{k+1}).$$

(Note: (1.2) is equivalent to the standard definition stating that the past and future are conditionally independent given the present.)

It is easy to show that such a Markov process P is also Markovian in the following spatial sense:

Let I be a finite interval of at least 3 integers with endpoints m, n where $m < n$. Let $I^o = \{k : m < k < n\}$ (the interior of I), and let $I^c = Z-I$ (the exterior of I). Then, relative to P, $\sigma(x(k), k \in I^o)$ and $\sigma(x(k), k \in I^c)$ are conditionally independent given $\sigma(x(m),x(n))$ on $\{x(m),x(n) \in S\}$.

The first type of homogeneity we consider relates to the spatial Markov property described above.

(1.3) DEFINITION. A Markov process is homogeneous if for every interval of at least 3 integers, I, the distribution of the process evaluated on I^o given the values (in S) at the 2 endpoints depends only on the length of I.

It turns out that (1.3) follows from the above definition restricted to intervals of just 3 integers. If I has just 3 integers then I^o simply represents the middle integer, and the conditional distribution of the process evaluated at the middle integer given the values (in S) at the endpoints we shall call the local characteristics in accord with the terminology from Markov random field theory. Thus the technical definition given in (2.1) will only deal with intervals of 3 integers, where we denote the local characteristics with ℓ.

The other 2 definitions of homogeneity considered here are temporal in nature. A process is called homogeneous in the forward time direction if for any integer n, the distribution of the process evaluated on $\{n+1, n+2,...\}$ given the value (in S) at time n is independent of n. A similar definition describes processes which are homogeneous in the backward time direction. A Markov process which is homogeneous in both time directions is called a two-sided Markov chain, and these have been studied extensively in [1]. We now briefly describe the main results.

In section 2 it is shown that if P is Markov and homogeneous and

$P(x(n) = i,\ x(n+1) = j) > 0\ \forall\ n \in Z,\ i,\ j \in S$, then there is a positive matrix p, not necessarily substochastic, a family of measures $\{\pi_n, n \in Z\}$, and a family of functions $\{h_n, n \in Z\}$, all strictly positive, so that

(1.4) $P(x(n) = i) = \pi_n(i)h_n(i)\ \forall\ n \in Z,\ i \in S$, and

(1.5) $P(x(n+k) = i_k : 0 \leq k \leq m) = \pi_n(i_0)\left[\prod_{k=0}^{m-1} p(i_k, i_{k+1})\right] h_{n+m}(i_m)$

whenever $m, n \in Z$, $m \geq 1$, and $i_0, i_1, \ldots i_m \in S$.

The argument is a modification of one found in 6 where Gibbs states are studied. Indeed, if we also demand that P be a probability and that $x(n) \in S\ \forall\ n \in Z$, P-a.s., then P is an example of a <u>Gibbs state specified by</u> p. With this as motivation we say that a process P is <u>specified by</u> p if there are non-negative families $\{\pi_n, n \in Z\}$ and $\{h_n, n \in Z\}$ so that (1.4) and (1.5) hold. It is shown that such a P is Markov and homogeneous; see (2.15). (In [6] it is shown that extreme Gibbs states are probability measures which satisfy (1.4) and (1.5) for positive p, π_n, and h_n.)

In (2.16)we state that the local characteristics for a process P specified by p are equal to $p(i,j)p(j,k)[p^2(i,k)]^{-1}$, which is quite similar to Theorem 1 of [6]. Indeed, in [6] it is mentioned that such a form for the local characteristics follows from the equivalence of homogeneous Markov random fields and nearest neighbor Gibbs states; also see [5]. The proof given here is easier becausewe started with the assumption that the process was Markovian in the traditional sense, whereas a general Markov random field need not be a Markov process.

Now let P be specified by p. Then $\forall\ n \in Z$ and $j \in S$ we have $\sum_{i \in S} \pi_n(i)p(i,j)h_{n+1}(j) = \sum_{i \in S} P(x(n) = i,\ x(n+1) = j) \leq P(x(n+1) = j) = \pi_{n+1}(j)h_{n+1}(j)$. Thus

(1.6) $\pi_n p(j) \leq \pi_{n+1}(j)$ whenever $h_{n+1}(j) \neq 0$.

Similarly,

(1.7) $ph_{n+1}(i) \leq h_n(i)$ whenever $\pi_n(i) \neq 0$.

(Note: We are using here standard notation concerning matrices, measures, and functions on S which are non-negative. Thus, if p is a matrix, π is a measure, and h is a function we have πp is a new measure with $\pi p(j) = \sum_{i \in S} \pi(i) p(i,j) \; \forall \; j$, and ph is a new function with $ph(i) = \sum_{j \in S} p(i,j) h(j) \; \forall \; i$.)

Conversely, given p, π_n, and $h_n \geq 0$ which satisfy (1.6) and (1.7), and one additional consistency hypothesis (see (2.20), it is possible to construct P satisfying (1.4) and (1.5); see (2.22) .

Again suppose P is specified by p. The condition that P be homogeneous in the forward time direction is essentially equivalent to the statement that $\forall \; i, j \in S$ the expression $h_n(i)[h_{n+1}(j)]^{-1}$ be independent of n; see (3.3) . Similarly, the condition that P be homogeneous in the backward time direction is essentially equivalent to the statement that $\forall \; i, j \in S$ the expression $\pi_n(i)[\pi_{n+1}(j)]^{-1}$ be independent of n; see (3.4) . Now, if P is homogeneous in both time directions (i.e. P is a two-sided Markov chain) and the forward one-step transition matrix is irreducible (plus one additional hypothesis; see (3.6)) then there is an integer $d \geq 1$ and real numbers $\sigma, \tau > 0$ so that $P \circ \theta_{nd}^{-1} = \rho^n P \; \forall \; n \in Z$ (here $\rho = \sigma\tau$) and whenever $P(x(n) = i) \neq 0$ then $\pi_{n+d}(i) = \sigma \pi_n(i)$ and $h_{n+d}(i) = \tau h_n(i)$; see (3.7) . In other words, in this case P exhibits a type of periodic behavior, and so do the sequences $\{\pi_n, n \in Z\}$, $\{h_n, n \in Z\}$ in the manner described above. (Note: In the case where P is a Gibbs state specified by p (see [6]), then it turns out that $d = \rho = 1$, and as such any Gibbs state specified by p which is a two-sided Markov chain must be translation invariant.)

This last result leads to a method of constructing a process specified by p which has the same type of periodic behavior as above, even if the irreducibility hypothesis on the one-step transition matrix is removed; see (3.22) . The essential idea is that the construction should only depend on the finite number of objects p, $\pi_0, \ldots, \pi_{d-1}, h_0, \ldots, h_{d-1}$. We end section 3 with 2 examples illustrating the above construction, the first of which having d = 2. The second, with d = 1, goes as follows: We are given a matrix $p \geq 0$, a measure $\pi \geq 0$, and a function $h \geq 0$, and 2 real numbers $\sigma, \tau > 0$ so that $\pi p \leq \sigma \pi$ and $ph \leq \tau^{-1} h$. It is shown that that (1.4) and (1.5) can be realized by letting $\pi_n = \sigma^n \pi \; \forall \; n \in Z$ and $h_n = \tau^n h \; \forall \; n \in Z$. Thus it is possible to simultaneously represent in a single process P a σ-excessive (relative to p) measure π and a τ^{-1}-excessive (relative to p) function h. It follows that then $P \circ \theta_n = \rho^n P \ \forall \ n \in Z$, where $\rho = \sigma\tau$; such a process is called quasi-stationary in [1]. (Note: The above discussion of construction is in the spirit of Markov random field theory in which one of the main questions is the existence of a field with given local characteristics; see [5].)

2. HOMOGENEOUS MARKOV PROCESSES.

The following definition follows the definition of homogeneity for Markov random fields as found in, e.g. [3].

(2.1) DEFINITION. Let P be a Markov process. P is called <u>homogeneous</u> if there exists $\ell : S^3 \to [0,1]$ so that $\forall \ i \in S$ and $n \in Z$, $P(x(n) = i \mid x(n-1), x(n+1)) = \ell(x(n-1), i, x(n+1))$ on $\{x(n-1),\ x(n+1) \in S\}$.

We now show that if all measures on basic cylinder sets are positive then a homogeneous Markov process must have a very nice form. The following proof is a modification of one found in [6].

(2.2) THEOREM. *Let P be a homogeneous Markov process. Let $\{p_n,\ n \in Z\}$ and ℓ be as in (1.1e) and (2.1) respectively. Assume that $\forall \ n \in Z$, and $i, j \in S$, $P(x(n) = i,\ x(n+1) = j) \neq 0$. Then there exist a strictly positive finite valued matrix p on S, a family $\{\pi_n,\ n \in Z\}$ of strictly positive sigma-finite measures on S, and a family $\{h_n,\ n \in Z\}$ of strictly positive functions (finite valued) on S so that (1.4) and (1.5) hold.*

PROOF. Fix $n \in Z$, $i, j \in S$. Then $P(x(n) = i,\ x(n+1) = j) = P(x(n) = i) p_n(i,j)$. Letting $\mu_n = P(x(n) = i) \ \forall \ n \in Z$, $i, j \in S$, it follows that $\mu_n > 0$ and $p_n > 0 \ \forall \ n \in Z$. (1.2) thus implies that $P(x(n+k) = i_k : 0 \leq k \leq m) \neq 0$ whenever $m, n \in Z$, $m \geq 1$, and $i_0, i_1, \ldots i_m \in S$. This in turn implies that $\ell > 0$.

Fix $i_0 \in S$. Then (1.2) and (2.1) imply that $\forall \ n \in Z$, $i, j \in S$,

(2.3) $\ell(i,j,i_0) = p_n(i,j) p_{n+1}(j,i_0) [p_n p_{n+1}(i,i_0)]^{-1}$ and

(2.4) $\ell(j,i_0,i_0) = p_{n+1}(j,i_0) p_{n+2}(i_0,i_0) [p_{n+1} p_{n+2}(j,i_0)]^{-1}$.

Dividing these expressions gives

(2.5) $\ell(i,j,i_0)[\ell(j,i_0,i_0)]^{-1}$
$= p_n(i,j)p_{n+1}p_{n+2}(j,i_0)[p_{n+2}(i_0,i_0)p_np_{n+1}(i,i_0)]^{-1}$.

Define a sequence of numbers $\{c_n, n \in Z\}$ and a sequence of functions $\{h_n, n \in Z\}$ by

(2.6) $c_0 = 1$, $c_n c_{n+1}^{-1} = p_{n+2}(i_0,i_0)^{-1}$ and

(2.7) $h_n(i) = c_n[p_np_{n+1}(i,i_0)]^{-1}$.

By (2.5) we now have

(2.8) $\ell(i,j,i_0)[\ell(j,i_0,i_0)]^{-1} = h_n(i)p_n(i,j)[h_{n+1}(j)]^{-1}$.

Thus if we define a matrix p by

(2.9) $p(i,j) = \ell(i,j,i_0)[\ell(j,i_0,i_0)]^{-1}$

then (2.8) implies

(2.10) $h_n(i)p_n(i,j) = p(i,j)h_{n+1}(j)$ $\forall$ $n \in Z$, and $i, j \in S$.

Now we define $\{\pi_n, n \in Z\}$ by

(2.11) $\pi_n(i) = P(x(n) = i)[h_n(i)]^{-1}$.

Clearly (1.4) holds. Also if $n \in Z$, and $i, j \in S$ then $P(x(n) = i, x(n+1) = j) = \pi_n(i)h_n(i)p_n(i,j) = \pi_n(i)p(i,j)h_{n+1}(j)$, and hence (1.5) holds if $m = 1$. An easy induction argument checks (1.5) for all $m \geq 1$, and the proof is complete.

REMARK. Let P, p, π_n, and h_n be as in (2.2). It follows that $P(x(n) = i, x(n+m) = j) = \pi_n(i)p^m(i,j)h_{n+m}(j)$, whenever $m, n \in Z$, $m \geq 1$, and $i, j \in S$. Thus each power of p is also finite valued. With this in mind it is possible to give a nice version of the local characteristics and the conditional distributions mentioned in (1.3). The easy proof is omitted.

(2.12) COROLLARY. Let P, p , π_n, and h_n be as in (2.2). Define, for m ≥ 1 an integer, $\ell_m : S^{m+2} \to [0,1]$ by

$$(2.13)\quad \ell_m(i_0,i_1,\ldots,i_m,i_{m+1}) = \left[\prod_{k=0}^{m} p(i_k,i_{k+1})\right]\left[p^{m+1}(i_0,i_{m+1})\right]^{-1}.$$

Then $\forall$ m, n $\in$ Z, m $\geq$ 1, and $i_1,\ldots,i_m \in S$ we have
$P(x(n+k) = i_k,\ 1 \leq k \leq m \mid x(n),\ x(n+m+1)) =$
$\ell_m(x(n),i_1,\ldots,i_m,x(n+m+1))$ on $\{x(n),\ x(n+m+1)\} \in S$. Also, $\ell = \ell_1$.

Theorem (2.2) leads to the following definition.

(2.14) DEFINITION. Let P be a process and p a non-negative finite valued matrix on S. Then we will say that P is specified by p if there exists a family $\{\pi_n,\ n \in Z\}$ of sigma-finite measures on S and a family $\{h_n,\ n \in Z\}$ of non-negative finite valued functions on S so that (1.4) and (1.5) hold. (Note: We do not require here that p, π_n, and h_n be strictly positive.)

It follows that if P is specified by p then P is Markov and homogeneous. The straightforward (and somewhat tedious)proof is left to the reader.

(2.15) THEOREM. Let P be specified by p as in (2.14). Then P is Markov and homogeneous.

Another consequence of P being specified by p is that P has the spatial Markov property mentioned in the introduction and that (1.3) holds in the following specific sense.

(2.16) THEOREM. Let P be specified by p as in (2.14). For m $\geq$ 1 define $\ell_m : S^{m+2} \to [0,1]$ by formula (2.13) if $0 < p^{m+1}(i_0,i_{m+1}) < \infty$ and let $\ell_m(i_0,i_1,\ldots,i_m,i_{m+1}) = 0$ otherwise. Fix m, n $\in$ Z , m $\geq$ 1, and let I = $\{k : n \leq k \leq n+m+1\}$. Letting I^o and I^c be as in the introduction we have that , relative to P, $\sigma(x(k),\ k \in I^o)$ and $\sigma(x(k),\ k \in I^c)$ are conditionally independent given $\sigma(x(n),\ x(n+m+1))$ on $\{x(n),\ x(n+m+1) \in S\}$. Further, $P(x(n+k) = i_k,\ 1 \leq k \leq m \mid x(n),\ x(n+m+1)) =$

$\ell_m(x(n),i_1,\ldots,i_m,x(n+m+1))$ _on_ $\{x(n),\ x(n+m+1) \in S\}$.

PROOF. The straightforward proof is left to the reader.

We conclude this section with a method of constructing processes specified by matrices. First we need a general result concerning the construction of Markov processes.

Suppose P is Markov as in (1.1e) . By (1.2) it follows that $\sum_{i \in S} P(x(n) = i)p_n(i,j) \leq P(x(n+1) = j)$. That is, if we define $\mu_n(i) = P(x(n) = i)$ then we have

(2.17) $\mu_n p_n \leq \mu_{n+1}\ \forall\ n \in Z.$

The following theorem states that (2.17) is also a sufficient condition for the construction of Markov processes. The proof is very similar to Theorem 3.1 of [1], and is omitted; in the case where each p_n is stochastic and $\mu_n p_n = \mu_{n+1}\ \forall\ n \in Z$, this follows from Theorem B of [2].

(2.18) THEOREM. _Let_ $\{\mu_n,\ n \in Z\}$ _be a family of sigma-finite measures on_ S, _and_ $p_n,\ n \in Z$ _a family of substochastic matrices on_ S _which satisfy_ (2.17). _Then there exists a Markov process_ P _so that_ $\mu_n(i) = P(x(n) = i)\ \forall\ n \in Z$ _and_ $i \in S$, _and_ (1.2) _holds_.

Now suppose P, p, π_n, and h_n are as in (2.14). Fix $n \in Z$ and $j \in S$. Then $\pi_n p(j)h_{n+1}(j) = \sum_{i \in S}\pi_n(i)p(i,j)h_{n+1}(j) = \sum_{i \in S} P(x(n) = i,\ x(n+1) = j) \leq P(x(n+1) = j) = \pi_{n+1}(j)h_{n+1}(j)$. Thus

(2.19) $\pi_n p(j) \leq \pi_{n+1}(j)$ whenever $h_{n+1}(j) \neq 0$.

Also, if $m, n \in Z$, $m \geq 1$, and $i_0,\ldots,i_m \in S$ then

(2.20) $\pi_n(i_0) \prod_{k=0}^{m-1} p(i_k,i_{k+1})\ h_{n+m}(i_m) \neq 0 \Rightarrow \pi_{n+k}(i_k)h_{n+k}(i_k) \neq 0$ for $0 \leq k \leq m$

since the first term in (2.20) is the same as $P(x(n+k) = i_k,\ 0 \leq k \leq m)$ and since for $0 \leq k \leq m$, $P(x(n+k) = i_k) = \pi_{n+k}(i_k)h_{n+k}(i_k)$.

Similar to the proof of (2.19) we can prove

(2.21) $ph_{n+1}(i) \leq h_n(i)$ whenever $\pi_n(i) \neq 0$.

Thus (2.19-21) are necessary conditions for the existence of a process satisfying (1.4) and (1.5). We now prove their sufficiency.

(2.22) THEOREM. (2.19-21) <u>imply there exists a process satisfying definition</u> (2.14).

PROOF. For $n \in Z$ define a matrix p_n on S by

(2.23) $p_n(i,j) = p(i,j)h_{n+1}(j)\, h_n(i)^{-1}$ if $\pi_n(i)h_n(i) \neq 0$, and $= 0$

otherwise.

Then (2.21) $\Rightarrow$ each p_n is substochastic.

Next fix $i, j \in S$ and $n \in Z$. If $\pi_n(i)h_n(i) \neq 0$ then $\pi_n(i)h_n(i)p_n(i,j) = \pi_n(i)p(i,j)h_{n+1}(j)$. The same conclusion holds if $\pi_n(i) = 0$. Finally, if $\pi_n(i) \neq 0$ and $h_n(i) = 0$, then (2.21) $\Rightarrow$ $p(i,j)h_{n+1}(j) = 0$. Thus we have

(2.24) $\pi_n(i)h_n(i)p_n(i,j) = \pi_n(i)p(i,j)h_{n+1}(j)\ \forall\ i, j \in S$ and $n \in Z$.

For $n \in Z$ and $j \in S$, $h_{n+1}(j) = 0 \Rightarrow \Sigma_{i\in S}\pi_n(i)h_n(i)p_n(i,j)$ $= \Sigma_{i\in S}\pi_n(i)p(i,j)h_{n+1}(j) = 0 = \pi_{n+1}(j)h_{n+1}(j)$, whereas $h_{n+1}(j) \neq 0 \Rightarrow$, by (2.19), $\Sigma_{i\in S}\pi_n(i)h_n(i)p_n(i,j) = \Sigma_{i\in S}\pi_n(i)p(i,j)h_{n+1}(j) =$ $\pi_n p(j)h_{n+1}(j) \leq \pi_{n+1}(j)h_{n+1}(j)$. Thus we have

(2.25) $\Sigma_{i\in S}\pi_n(i)h_n(i)p_n(i,j) \leq \pi_{n+1}(j)h_{n+1}(j)\ \forall\ j \in S$ and $n \in Z$.

By (2.18) there exists a Markov process P such that $P(x(n) = i)$ $= \pi_n(i)h_n(i)\ \forall\ i \in S$ and $n \in Z$, and which satisfies (1.2).

We will now show by induction on m that (1.5) holds. The case m = 1 is precisely (2.24). Fix $m \geq 1$ and suppose (1.5) holds for all choices of n, $i_0, i_1, \ldots, i_m$. Fix $n \in Z$ and $i_0, \ldots, i_m, i_{m+1} \in S$. By (1.2) $P(x(n+k) = i_k,\ 0 \leq k \leq m+1) = P(x(n+k) = i_k,\ 0 \leq k \leq m)p_{n+m}(i_m, i_{m+1})$, which, by the induction hypothesis,

$= \pi_n(i_0)\left[\prod_{k=0}^{m-1} p(i_k,i_{k+1})\right] h_{n+m}(i_m)p_{n+m}(i_m,i_{m+1})$. If $\pi_{n+m}(i_m)h_{n+m}(i_m) \neq 0$

then (2.23) $\Rightarrow P(x(n+k) = i_k,\ 0 \leq k \leq m+1)$

$= \pi_n(i_0)\left[\prod_{k=0}^{m} p(i_k,i_{k+1})\right] h_{n+m+1}(i_{m+1})$. Also, if $\pi_{n+m}(i_m)h_{n+m}(i_m) = 0$

then $P(x(n+k) = i_k,\ 0 \leq k \leq m+1) \leq P(x(n+m) = i_m) = \pi_{n+m}(i_m)h_{n+m}(i_m)$

$= 0$ and, by (2.20), $\pi_n(i_0)\left[\prod_{k=0}^{m} p(i_k,i_{k+1})\right] h_{n+m+1}(i_{m+1}) = 0$. Thus (1.5)

holds for m+1, and the induction argument is complete.

(2.26) REMARKS. We briefly describe a few ways in which (2.20) can be eliminated as a hypothesis in theorem (2.22). We first observe that if $\pi_n p \leq \pi_{n+1}\ \forall\, n \in Z$ then it easily follows by induction that $\forall\, m, n \in Z$ and $m \geq 1$ we have $\pi_n p^m \leq \pi_{n+m}$. Similarly if $ph_n \leq h_{n-1}\ \forall\, n \in Z$ then $p^m h_n \leq h_{n-m}\ \forall\, m, n \in Z$ and $m \geq 1$. Using these observations we may replace the hypotheses in (2.22) by either of the following 3 alternatives:

(a) (2.19), (2.21), and the condition that $h_n > 0\ \forall\, n \in Z$.

(b) (2.19), (2.21), and the condition that $\pi_n > 0\ \forall\, n \in Z$.

(c) $\pi_n p \leq \pi_{n+1}$, and $ph_n \leq h_{n-1}\ \forall\, n \in Z$.

We omit the relatively easy details which show that either (a), (b), or (c) implies (2.19-21). The point is that these alternative hypotheses are less cumbersome.

3. HOMOGENEOUS MARKOV PROCESSES WHICH ARE TWO-SIDED MARKOV CHAINS.

We begin this section with precise definitions , with appropriate notation, of temporal homogeneity.

(3.1) DEFINITIONS. Let P be a Markov process.

(a) P is <u>homogeneous in the forward time direction</u> if there exists a

substochastic matrix q on S such that $\forall$ n $\in$ Z, and i, j $\in$ S,
$P(x(n) = i, x(n+1) = j) = P(x(n) = i)q(i,j)$.
(b) P is <u>homogeneous in the backward time direction</u> if there exists a substochastic matrix r on S such that $\forall$ n $\in$ Z, and i, j $\in$ S,
$P(x(n-1) = j, x(n) = i) = P(x(n) = i)r(i,j)$.
(c) P is called a <u>two-sided Markov chain</u> if P is homogeneous in both time directions.

Suppose P is specified by p as in (2.14). Further assume that P is homogeneous in the forward time direction, and let q be as in (3.1a). Then

(3.2) $\pi_n(i)p(i,j)h_{n+1}(j) = \pi_n(i)h_n(i)q(i,j)$ $\forall$ n $\in$ Z and i, j $\in$ S.

It follows that, by some simple manipulations,

(3.3) $h_m(i)h_{n+1}(j) = h_n(i)h_{m+1}(j)$ whenever m, n $\in$ Z, i, j $\in$ S, and

$\pi_n(i)\pi_m(i)p(i,j) \neq 0$.

Similarly we can show that if P is homogeneous in the backward time direction then

(3.4) $\pi_n(i)\pi_{m+1}(j) = \pi_m(i)\pi_{n+1}(j)$ whenever m, n $\in$ Z, i, j $\in$ S, and

$p(i,j)h_{n+1}(j)h_{m+1}(j) \neq 0$.

(3.5) PROPOSITION. <u>Let</u> P <u>be specified by</u> p <u>as in</u> (2.14).
(i) (3.3) $\Rightarrow$ P <u>is homogeneous in the forward time direction.</u>
(ii) (3.4) $\Rightarrow$ P <u>is homogeneous in the backward time direction.</u>

PROOF. We only prove (i) as (ii) is quite similar. Thus suppose (3.3). Define matrix q on S by : $q(i,j) = p(i,j)h_{n+1}(j)[h_n(i)]^{-1}$ if $\pi_n(i)h_n(i) \neq 0$ for some n, and = 0 otherwise. Then (3.3) $\Rightarrow$ q is well-defined, and (2.21) $\Rightarrow$ q is substochastic. Fix n $\in$ Z, and i, j $\in$ S. If $\pi_n(i)h_n(i) \neq 0$ then by definition of q, $P(x(n) = i, x(n+1) = j)$ $= \pi_n(i)p(i,j)h_{n+1}(j) = \pi_n(i)h_n(i)q(i,j)$. If $\pi_n(i)h_n(i) = 0$ then $P(x(n) = i, x(n+1) = j) = 0 = \pi_n(i)h_n(i)q(i,j)$. Thus in any case

$P(x(n) = i, x(n+1) = j) = P(x(n) = i)q(i,j)$, and we are done.

In [1] the assumption of irreducibility of the forward one-step transition matrix of a two-sided Markov chain was shown to be sufficient for the main structure theorem; Theorem 4.2 of [1] contains the precise statement. Here we make a similar type of communication assumption.

(3.6) HYPOTHESES. Let P be a process.

(a) There exisits $m \geq 1$ such that $\forall\, i \in S$ there exists $n \in Z$ such that $P(x(n) = i, x(n+m) = i) \neq 0$.

(b) $\forall\, i, j \in S$, there exist $m, n \in Z$, $m \geq 1$ such that $P(x(n) = i, x(n+m) = j) \neq 0$.

REMARK. Loosely, (3.6a) means there is some $m \geq 1$ so that for every i it is possible to go from i to i in m steps. (3.6b) means that $\forall\, i, j$ it is possible to go from i to j in at least one step.

(3.7) THEOREM. *Let P be specified by p as in* (2.14), *and assume P is a two-sided Markov chain. Further suppose* (3.6). *Then there exists* $d \in Z$, $d \geq 1$, *and positive real numbers* σ, τ *so that whenever* $P(x(n) = i) \neq 0$ *then* $\pi_{n+d}(i) = \sigma\pi_n(i)$ *and* $h_{n+d}(i) = \tau h_n(i)$, *and* $P \circ \theta_{nd}^{-1} = \rho^n P\ \forall\, n \in Z$ *where* $\rho = \sigma\tau$.

PROOF. Let q be as in (3.1a). It follows that for $n \in Z$, $\mu_n q \leq \mu_{n+1}$ where $\mu_n(i) = P(x(n) = i)\ \forall\, n \in Z$ and $i \in S$. It follows by induction that $\mu_n q^m \leq \mu_{n+m}\ \forall\, m, n \in Z$, $m \geq 1$. Also, since q is the forward one-step transition matrix for P we have,

$$(3.8)\quad P(x(n+k) = i_k,\ 0 \leq k \leq m) = \mu_n(i_0)\prod_{k=0}^{m-1} q(i_k, i_{k+1}) \text{ whenever}$$

$m, n \in Z$, $m\quad 1$, and $i_0, \ldots, i_m \in S$.

Combining (3.6) and (3.8) it follows that q is irreducible (i.e. $\forall\, i, j \in S$ there exists $m \in Z$, $m \geq 1$, such that $q^m(i,j) \neq 0$) and $T \equiv \{m \geq 1 : q^m(i,i) \neq 0\ \forall\, i \in S\} \neq \emptyset$. Note that if $m, n \in T$ then $q^{m+n}(i,i) \geq q^m(i,i)q^n(i,i) \neq 0\ \forall\, i \in S$, and thus $m+n \in T$. That is, T is closed under addition. Let d be the greatest common divisor of T.

By [4], Lemma 1-66, there exists $K \in Z$, $K \geq 1$, so that $kd \in T$ whenever $k \geq K$.

By [1], Theorem 4.2, $\forall\, n \in Z$ and $i \in S$, $\mu_n(i) = 0 \Leftrightarrow \mu_{n+d}(i) = 0$. Also if $i, j \in S$, $m, n \in Z$, and $m \geq 1$, then (3.8) $\Rightarrow$

$$(3.9)\quad \pi_n(i)p^m(i,j)h_{n+m}(j) = \pi_n(i)h_n(i)q^m(i,j).$$

Thus for $n \in Z$ and $i \in S$ we have that $\pi_n(i)p^{Kd}(i,i)h_{n+Kd}(i) = \pi_n(i)h_n(i)q^{Kd}(i,i)$ and $\pi_n(i)p^{(K+1)d}(i,i)h_{n+(K+1)d}(i) = \pi_n(i)h_n(i)q^{(K+1)d}(i,i)$. Thus, given i, we may select, according to (3.6), $n \in Z$ with $\pi_n(i)h_n(i) \neq 0$, and conclude that $0 < p^{Kd}(i,i)$, $p^{(K+1)d}(i,i) < \infty$. We may then define

$$(3.10)\quad \tau_i = q^{(K+1)d}(i,i)p^{Kd}(i,i)\left[q^{Kd}(i,i)p^{(K+1)d}(i,i)\right]^{-1}.$$

It now follows that

$$(3.11)\quad h_{n+d}(i) = \tau_i h_n(i) \text{ if } \mu_n(i) \neq 0.$$

Now let r be as in (3.1b). A similar argument as for q leads to

$$(3.12)\quad \pi_n(i)h_n(i)q^m(i,j) = \pi_{n+m}(j)h_{n+m}(j)r^m(j,i) \text{ whenever } m, n \in Z,$$

$m \geq 1$, and $i, j \in S$.

(3.6) $\Rightarrow \{m \geq 1 : q^m(i,j) \neq 0\} = \{m \geq 1 : r^m(j,i) \neq 0\}$. Letting T be as in the second paragraph of this proof, it follows that $T = \{m \geq 1 : r^m(i,i) \neq 0 \ \forall\, i \in S\}$. With d as before a similar argument as for q leads to the fact that $\forall\, i \in S$, $r^{Kd}(i,i)$, $r^{(K+1)d}(i,i) \neq 0$ and thus we may define

$$(3.13)\quad \sigma_i = r^{Kd}(i,i)p^{(K+1)d}(i,i)\left[r^{(K+1)d}(i,i)p^{Kd}(i,i)\right]^{-1}.$$

Similarly, as for q, it follows that

$$(3.14)\quad \pi_{n+d}(i) = \sigma_i \pi_n(i) \text{ whenever } \mu_n(i) \neq 0.$$

Fix $i, j \in S$, and choose $m, n \in Z$, $m \geq 1$, so that

$P(x(n) = i, x(n+m) = j) \neq 0$, according to (3.6). Thus by (3.11) and (3.14) we now have

$$(3.15)\quad \sigma_i \tau_i \pi_{n-d}(i) h_{n-d}(i) q^m(i,j) = \pi_n(i) h_n(i) q^m(i,j)$$

$$= \pi_n(i) p^m(i,j) h_{n+m}(j) = \sigma_i \pi_{n-d}(i) p^m(i,j) \tau_j h_{n+m-d}(j).$$

But $\pi_{n-d}(i) h_{n-d}(i) q^m(i,j) = \pi_{n-d}(i) p^m(i,j) h_{n+m-d}(j)$ and thus (3.15) $\Rightarrow \sigma_i \tau_i = \sigma_i \tau_j \Rightarrow \tau_i = \tau_j$. A similar argument involving r $\Rightarrow$ $\sigma_i = \sigma_j \; \forall \; i, j$. In summary, there exists $\sigma, \tau > 0$ so that

$$(3.16)\quad \pi_{n+d}(i) = \sigma \pi_n(i), \text{ and } h_{n+d}(i) = \tau h_n(i) \text{ whenever } \mu_n(i) \neq 0.$$

Letting $\rho = \sigma\tau$ then (3.16) $\Rightarrow$

$$(3.17)\quad \mu_{n+d} = \rho \mu_n \; \forall \; n \in Z.$$

An easy monotone class argument completes the argument.

(3.18) REMARKS.

(a) Let P be as in the preceding theorem. (3.10) and (3.13) $\Rightarrow$ $\rho = r^{Kd}(i,i) q^{(K+1)d}(i,i) [r^{(K+1)d}(i,i) q^{Kd}(i,i)]^{-1}$ which accords with formula (4.4) of [1]. Thus the constant ρ depends only on the forward and backward transitions whereas the constants σ, τ also depend on the (possibly non-substochastic) matrix p.

(b) Again let P be as in (3.7), and further assume $P(x(n) = i, x(n+1) = j) \neq 0 \; \forall \; n \in Z$ and $i, j \in S$; this represents a strengthening of (3.6). This easily implies $d = 1$. Setting $\pi = \pi_0$ and $h = h_0$, then $\pi_n = \sigma^n \pi$ and $h_n = \tau^n h \; \forall \; n \in Z$. This implies $h(i)q(i,j) = \tau p(i,j)h(j)$ and $\sigma\pi(j)r(j,i) = \pi(i)p(i,j)$. Thus the forward transition q is a "path transform" of p and the backward transition r is a "dual" of p. Note that if $\sigma = \tau = 1$ and p is substochastic, then the above conditions imply that q is the h-path transform of p and r is in duality with p with respect to π; see [4].

The rest of this section concerns conditions for the construction of two-sided Markov chains specified by a matrix. The advantage we will gain here over a general result such as (2.22) is that the construction we now consider only depends on a finite number of objects to begin with.

More specifically, if P is as in (3.7), then the result of (3.7) implies that P depends on σ, τ, d, $\pi_0, \ldots, \pi_{d-1}$, $h_0, \ldots, h_{d-1}$. We now specify some necessary conditions which will turn out to be the conditions sufficient to carry out our construction.

Letting P be as in (3.7) it follows simply that $\pi_{n+d}(i)p(i,j)h_{n+d+1}(j) = P(x(n+d) = i,\ x(n+d+1) = j) = \rho P(x(n) = i,\ x(n+1) = j) = \rho\pi_n(i)p(i,j)h_{n+1}(j)$. Thus

(3.19) $\pi_n(i)h_{n+1}(j) = 0 \Leftrightarrow \pi_{n+d}(i)h_{n+d+1}(j)$ whenever $p(i,j) \neq 0$.

Also, for easy reference, we record here two other necessary conditions.

(3.20) $\pi_n(i)h_n(i) = 0 \Leftrightarrow \pi_{n+d}(i)h_{n+d}(i) = 0$, $\forall\, n \in Z$ and $i \in S$.

(3.21) $\pi_{n+d}(i) = \sigma\pi_n(i)$, and $h_{n+d}(i) = \tau h_n(i)$ whenever $\pi_n(i)h_n(i) \neq 0$.

We have everything required to state the following construction theorem.

(3.22) THEOREM. *Suppose given the following objects*: (1) *a non-negative finite valued matrix* p *on* S; (2) *an integer* $d \geq 1$; *positive real numbers* σ, τ ; (3) *a family of sigma-finite measures on* S, $\{\pi_n,\ n \in Z\}$; (4) *a family of non-negative finite valued functions on* S, $\{h_n,\ n \in Z\}$.

Further assume

(i) (2.20), (3.19-21).

(ii) (3.3,4) *hold whenever* $0 \leq m, n \leq d-1$.

(iii) (2.19,21) *hold whenever* $0 \leq n \leq d-1$.

Then there exists a homogeneous Markov process P *specified by* p *such that* (1.4,5) *hold*, *and such that* $P \circ \theta_{nd}^{-1} = \rho^n P\ \forall\, n \in Z$ *where* $\rho = \sigma\tau$.

PROOF. As before let $\mu_n(i) = \pi_n(i)h_n(i)$. Define a matrix q on S as follows: If there exists $0 \leq n \leq d-1$ so that $\mu_n(i) \neq 0$ then let $q(i,j) = p(i,j)h_{n+1}(j)[h_n(i)]^{-1}$, and otherwise let $q(i,j) = 0$. By (ii) in the statement of this theorem, q is well-defined. Also by (iii) in the statement of this theorem, q is substochastic.

Similiarly define a matrix r on S as follows: If there exists $0 \leq n \leq d-1$ with $\mu_{n+1}(j) \neq 0$ then let $r(j,i) = \pi_n(i)p(i,j)[\pi_{n+1}(j)]^{-1}$, and otherwise let $r(j,i) = 0$. As with q, (ii) and (iii) in the statement of this theorem imply r is well-defined and substochastic.

Fix $i, j \in S$ and $0 \leq n \leq d-1$. If $\mu_n(i) \neq 0$ then, by definition, $\pi_n(i)h_n(i)q(i,j) = \pi_n(i)p(i,j)h_{n+1}(j)$. Clearly this equality also holds if $\pi_n(i) = 0$. Finally suppose $\pi_n(i) \neq 0$ and $h_n(i) = 0$. By (iii) in the statement $ph_{n+1}(i) \leq h_n(i) = 0$, and thus $p(i,j)h_{n+1}(j) = 0$, and the equality holds in this case also. In summary we have

(3.23) $\pi_n(i)h_n(i)q(i,j) = \pi_n(i)p(i,j)h_{n+1}(j)$ whenever $i, j \in S$ and

$0 \leq n \leq d-1$.

A similar argument gives that

(3.24) $\pi_n(i)p(i,j)h_{n+1}(j) = \pi_{n+1}(j)h_{n+1}(j)r(j,i)$ whenever $i, j \in S$ and

$0 \leq n \leq d-1$.

We now extend (3.23) and (3.24) to all $n \in Z$. Fix $n \in Z$ and $i, j \in S$. Choose $0 \leq r \leq d-1$ and $m \in Z$ so that $n = r+md$. Then (3.20) and (3.21) $\Rightarrow$ $\pi_n(i)h_n(i)q(i,j) = \rho^m \pi_r(i)h_r(i)q(i,j)$, which, by (3.23), $= \rho^m \pi_r(i)p(i,j)h_{r+1}(j)$. There are three possibilities. Firstly, suppose $\pi_r(i)p(i,j)h_{r+1}(j) \neq 0$. By (iii) of the statement we have that $\mu_r(i), \mu_{r+1}(j) \neq 0$. Thus (3.20) and (3.21) $\Rightarrow$ $\pi_n(i) = \sigma^m \pi_r(i)$, $h_{n+1}(j) = \tau^m h_{r+1}(j)$, and $\pi_n(i)h_n(i)q(i,j) = \pi_n(i)p(i,j)h_{n+1}(j)$. Secondly , suppose $p(i,j) \neq 0$ and $\pi_r(i)h_{r+1}(j) = 0$. Then (3.19) $\Rightarrow$ $\pi_n(i)h_{n+1}(j) = 0$ and hence $\pi_n(i)h_n(i)q(i,j) = 0 = \pi_n(i)p(i,j)h_{n+1}(j)$. Finally, suppose $p(i,j) = 0$. Then clearly $\pi_n(i)h_n(i)q(i,j) = 0$ $= \pi_n(i)p(i,j)h_{n+1}(j)$. In summary we have shown that

(3.25) $\pi_n(i)p(i,j)h_{n+1}(j) = \pi_n(i)h_n(i)q(i,j)$ $\forall$ $i, j \in S$ and $n \in Z$.

A similar argument gives

(3.26) $\pi_n(i)p(i,j)h_{n+1}(j) = \pi_{n+1}(j)h_{n+1}(j)r(j,i)\ \forall\ i,\ j \in S$ and $n \in Z$.

Since q and r are substochastic then (3.25,26) $\Rightarrow$ (2.19,21). Since (2.20) is part of the hypothesis then (2.22) implies the existence of a process P satisfying definition (2.14). Also, according to definition (3.1) , (3.25,26) $\Rightarrow$ P is a two-sided Markov chain. Finally (3.20,21) $= \mu_{m+nd} = \rho^n \mu_m\ \forall\ m,\ n \in Z$, and an easy monotone class argument gives that $P \circ \theta_{nd}^{-1} = \rho^n P\ \forall\ n \in Z$.

The follwing corollary provides for a case where the conditions of theorem (3.22) can be checked with knowledge of only $\pi_0, \pi_1, \ldots, \pi_{d-1}, h_0, h_1, \ldots, h_{d-1}$.

(3.27) COROLLARY. Suppose given p, d, σ, τ, as in (3.22). Also suppose given sigma-finite measures $\pi_0, \pi_1, \ldots, \pi_{d-1}$ and non-negative finite valued functions $h_0, h_1, \ldots, h_{d-1}$ with $\pi_n, h_n > 0\ \forall\ 0 \leq n \leq d-1$. Define $\{\pi_n, n \in Z\}$ via the formula $\pi_{n+d} = \sigma \pi_n\ \forall\ n \in Z$, and define $\{h_n, n \in Z\}$ via the formula $h_{n+d} = \tau h_n\ \forall\ n \in Z$, and assume $h_m(i)h_{n+1}(j) = h_n(i)h_{m+1}(j)$ and $\pi_n(i)\pi_{m+1}(j) = \pi_m(i)\pi_{n+1}(j)$ whenever $p(i,j) \neq 0$ and $0 \leq m, n \leq d-1$. Then there exists a process P satisfying the same conclusion as in (3.22), provided $\pi_n p \leq \pi_{n+1}$ and $ph_{n+1} \leq h_n$ whenever $0 \leq n \leq d-1$.

PROOF. It is a simple matter to check all the hypotheses of theorem (3.22).

(3.28) EXAMPLE. Let $S = \{1,2,3\}$, $p = \begin{bmatrix} 0 & 1 & 1 \\ 1 & 0 & 0 \\ 1 & 0 & 0 \end{bmatrix}$, $d = 2$, $\sigma = 2$, and $\tau = 2^{-1}$. Define $\{\pi_n, n \in Z\}$ and $\{h_n, n \in Z\}$ by the following rules (Note: In what fbllows k, m $\in$ Z and, by convention, we represent measures by rows and functions by columns.):

(a) $n = 2k \Rightarrow \pi_n = (2^{k+1}, 0, 0)$.
(b) $n = 2k+1 \Rightarrow \pi_n = (0, 2^{k+1}, 2^{k+1})$.
(c) $h_0^T = (1,2,2)$; here T denotes transpose.
(d) $n = 2k,\ k \neq 0, \Rightarrow h_n^T = (2^{-k}, 0, 0)$.
(e) $n = 4m+1 \Rightarrow h_n^T = (2^{1-2m}, 2^{-(1+2m)}, 2^{-(1+2m)})$.
(f) $n = 4m+3 \Rightarrow h_n^T = (0, 2^{-(2+2m)}, 2^{-(2+2m)})$.

We omit the details, but it can be shown that the hypotheses of (3.22) are satisfied. In this example we have that $\pi_{n+d} = \sigma\pi_n$ while it is not true that $h_{n+d} = \tau h_n$ $\forall n$. However (3.21) does hold. Also $\pi_n p = \pi_{n+1}$ but $ph_n \leq h_{n-1}$ if and only if n is even and nonzero, n=1, or $n \equiv 3 \bmod 4$. Nevertheless (2.21) holds. Thus this example serves to show that the various conditions in (3.22) cannot be simplified.

In this example it follows that $P \circ \theta_{2n}^{-1} = P$ since $\rho = 1$ and d = 2; thus P is periodic with period 2. Letting μ_n be the distribution of x(n) under P it is easy to see that $\mu_n = (2,0,0)$ if n is even, $\mu_n = (0,1,1)$ if n is odd, and $q = r = \begin{bmatrix} 0 & 2^{-1} & 2^{-1} \\ 1 & 0 & 0 \\ 1 & 0 & 0 \end{bmatrix}$. Since both q and r are stochastic then, P a.e., $x(n) \in S$ $\forall n \in Z$, i.e. the process is alive for all time. Thus the total mass is equal to $P(x(n) \in S) = \mu_n(1) + \mu_n(2) + \mu_n(3) = 2$.

We make one final remark concerning this example. It is easy to show that the communication hypotheses (3.6) hold and that $\{m \geq 1 : \forall\, i$ there exists $n \in Z$ such that $P(x(n) = i, x(n+m) = i) \neq 0\}$ is the set of positive even integers, and thus this example satisfies the hypotheses of (3.7).

(3.29) EXAMPLE. Let p be a non-negative finite valued matrix, π a sigma-finite measure, h a non-negative finite valued function, and σ, τ two positive real numbers such that $\pi p \leq \sigma\pi$ and $ph \leq \tau^{-1}h$. For $n \in Z$ define $\pi_n = \sigma^n \pi$ and $h_n = \tau^n h$. We omit the details, but it is easy to show that the hypotheses of (3.22) hold for d = 1.

Since $P \circ \theta_n^{-1} = \rho^n P$, then P is quasi-stationary; see [1]. Actually, it is possible to apply theorem 5.3 of [1] for the construction of P. Observe that P provides a way of simultaneously representing a σ- excessive (relative to p) measure π and a τ^{-1}-excessive function (relative to p) h.

REFERENCES.

[1] B.W. Atkinson. Two-sided Markov chains. Ann. Probab. Vol. 14, No. 2, 459-479, 1986.

[2] E.B. Dynkin. Boundary theory of Markov processes (the discrete case). Russian Math. Surveys Vol. 24, No. 2, 1-42, 1969.

[3] D. Griffeath. Introduction to random fields. Denumerable Markov Chains (by J.G. Kemeny, J.L. Snell, and A.W. Knapp). Springer-Verlag, New York, (1976).

[4] J.G. Kemeny, J.L. Snell, and A.W. Knapp. Denumerable Markov Chains. Springer-Verlag, New York, 1976.

[5] R. Kindermann and J.L. Snell. Markov Random Fields and their Applications. AMS series in Contemporary Mathematics, AMS, Providence, 1980.

[6] F. Spitzer. Phase transition in one-dimensional nearest neighbor systems. J. Funct. Anal. Vol. 20, 240-255, 1975.

Bruce W. Atkinson
Department of Mathematics
Palm Beach Atlantic College
1101 S. Olive Ave.
West Palm Beach, Florida 33401
U.S.A.

REGULARITY AND THE DOOB-MEYER DECOMPOSITION OF ABSTRACT QUASIMARTINGALES

by

J.K. BROOKS and N. DINCULEANU

Introduction.

In this paper we shall examine the structure of abstract quasi-martingales taking their values in a Banach space E. We are especially interested in the existence of cadlag modifications and in the Doob-Meyer decomposition of such quasimartingales. The motivation of this inquiry was twofold. First, the problems concerning regularity and the Doob-Meyer decomposition of quasimartingales arose naturally in the attempt to develop a stochastic integration theory for Banach space valued processes. The development of this abstract stochastic integral was, in turn, motivated by Kunita's theory for Hilbert space valued processes [9] and subsequent extensions, using the vector measure approach, by Pellaumail [16], Métivier and Pellaumail [14], Métivier [12] and Kussmaul [10] and [11]. Secondly, we wished to improve on the existing results on the regularity and structure of quasimartingales, as presented by the above mentioned authors.

Here is a presentation of the main results of this paper. Let $X:R_+ \times \Omega \to E$ be a (not necessarily cadlag) quasimartingale.

1.) Assume E has the Radon Nikodym Property (RNP). It turns out, surprisingly, that X is strongly cadlag if and only if X is weakly cadlag (theorem 4.2). Further, X is cadlag if and only if X has a Doob-Meyer decomposition X=M+A where M is a cadlag local martingale and A is a predictable, cadlag process with finite variation and $A_0 = 0$ (theorem 4.2). This decomposition is unique up to an evanescent set.

2.) Assume E has the RNP and X is a quasimartingale. Then X has a cadlag modification if and only if X satisfies the following regularity condition

(R) $\lim_{t \downarrow s} E(1_F X_t) = E(1_F X_s)$ for $s<\infty$ and $F \in \mathscr{F}_s$;

or even the following weak regularity condition:

(WR) $\lim_{t \downarrow s} \langle E(1_F X_t), z\rangle = \langle E(1_F X_s), z\rangle$ for $s<\infty, F \in \mathscr{F}_s$

and z in a subset Z of the dual E^*, which is norming for E. This is proved in theorem 4.1 which gives 13 equivalent conditions for the existence of a cadlag modification, such as: right continuity in probability; right continuity in the mean, that is $\lim_{t \downarrow s} X_t = X_s$ strongly in $L^1_E(P)$; right continuity weakly in $L^1_E(P)$; any of the above conditions for the scalar quasimartingales $\langle X,z\rangle$ with $z \in Z$. This result is new even in the scalar case.

As a corollary, (theorem 4.3) we give a characterization of the E-valued measure μ which is associated with a quasimartingale X by the equality

$$\mu((s,t]\times F) = E(1_F(X_t - X_s)), \text{ for } s<\infty \text{ and } F \in \mathcal{F}_s.$$

3.) Without any assumption on the Banach space E, we give the following results concerning the existence of a cadlag modification of X, in terms of its right limit X_+ along the rationals (theorems 5.2-5.5):

3a.) If X is a local martingale, then X_+ exists a.s., X_+ is a local martingale and a cadlag modification of X. We use the fact that this is true for martingales [2].

3b.) If X is a quasimartingale with separable range and if X_+ exists a.s., then X_+ is a right continuous modification of X if and only if the regularity condition (R) holds. If X is real valued, then X_+ and X_- exist a.s.

3c.) If X has integrable variation, then X_+ (which exists everywhere) is a cadlag modification of X if and only if condition (R) holds.

4.) Assume again that E has the RNP. Then X is a quasimartingale of class (D) and satisfies condition (R) if and only if X has a Doob-Meyer decomposition X=M+A with M a (not necessarily right continuous) martingale of class (D) and A a predictable, cadlag process with integrable variation and $A_0=0$ (theorems 6.1-6.4).

Conditions (D) and (R) together are also equivalent to the associated measure μ_X being σ-additive and with bounded variation.

Various properties concerning the concepts used in this paper will be stated in the Appendix and sections in this Appendix will be referred to as A.1, A.2, etc.

The proof of theorem 4.1 is broken into steps, and some independent lemmas will be given in the course of the proof (Lemma I, Lemma II, etc.).

1. Notation.

In this section, we shall present notation and definitions which will be used to prove the main results in the next sections.

Throughout the paper, $(\Omega,\mathcal{F},P)$ is a probability space, $(\mathcal{F}_t)_{t\in R_+}$ is a filtration satisfying the usual conditions, E is a Banach space with norm $\|\cdot\|$, and dual space E^*, and $(X_t)_{t\in R_+}$ is an E-valued, adapted process with $X_t \in L^1_E(P)$, for each $t \geq 0$. We shall always consider X to be extended to ∞, with $X_\infty = 0$. If X has a limit at ∞, it will be denoted by $X_{\infty-}$.

For every α, with $0 < \alpha \leq \infty$, we shall denote by $\mathcal{A}(0,\alpha]$, the ring generated by the predictable rectangles of the form $(s,t]\times F$, with $0 \leq s < t \leq \alpha$ and $F \in \mathcal{F}_s$, and by $\mathcal{A}(0,\alpha)$, the union of the rings $\mathcal{A}(0,\beta]$, with $\beta < \alpha$. In particular, $\mathcal{A}(0,\infty)$ consists of all finite, disjoint unions of bounded predictable rectangles $(s,t]\times F$ with $t<\infty$, and this ring generates the σ-algebra $\mathcal{P}$ of predictable subsets of $(0,\infty)\times\Omega$. Note that $\mathcal{A}(0,\alpha]$ contains the stochastic intervals $(S,T]$, where S and T are simple stopping times bounded by α. We notice that the above rings do not contain subsets of $\{0\}\times\Omega$. In fact, these sets have no role in problems of regularity.

2. The Doleans function.

For each predictable rectangle $(s,t]\times F$, we define

$$\mu_X((s,t]\times F) = E(1_F(X_t - X_s)).$$

In particular, $\mu_X((s,\infty]\times F) = -E(1_F X_s)$.

Note that μ_X is finitely additive on the semi-ring of predictable rectangles, and thus it can be extended uniquely to an E-valued, finitely additive measure on the algebra $\mathscr{A}(0,\infty]$; this extension will still be denoted by μ_X. It is called the Doléans function of the process X. This function is invaluable in the study of quasimartingales. An immediate consequence of the definition of μ_X is the equivalence of the following two conditions:

(*) $\quad \lim_{t\downarrow s} E(1_F X_t) = E(1_F X_s)$;

(**) $\quad \lim_{t\downarrow s} \mu_X((s,t]\times F) = 0$.

As we mentioned previously, we refer to the regularity condition above as condition (R). Other properties of μ_X will be stated in section A.1.

3. The mean variation of X and quasimartingales.

For any α with $0 \leq \alpha \leq \infty$, we define the mean variation of X on $(0,\alpha]$ to be the number

$$\mathrm{Var}_X(0,\alpha] = \sup \Sigma_i \| E((X_{t_{i+1}} - X_{t_i}) \mid \mathscr{F}_{t_i}) \|_1 \leq \infty,$$

where the supremum is taken over all finite partitions

$0 < t_1 < \ldots < t_n = \alpha$. The mean variation $Var_X(0,\alpha)$ of X on $(0,\alpha)$ is defined similarly by taking $t_n < \alpha$.

It is important to be able to compute $Var_X(0,\alpha]$ by taking the supremum over different sets, for example, random partitions. Results concerning different representations of $Var_X(0,\alpha]$ are presented in A.2, along with the relationship between $Var_X(0,\alpha]$ and $|\mu_X|$, the variation of μ_X.

We say that X is a quasimartingale on $(0,\alpha]$, or on $(0,\alpha)$, if $Var_X(0,\alpha] < \infty$ or $Var_X(0,\alpha) < \infty$ respectively. We shall be interested in three types of quasimartingales, namely those on $(0,\infty]$, on $(0,\infty)$, and those on every bounded interval $(0,\alpha]$. Elementary facts concerning quasimartingales, along with some convergence properties are given in A.3 and A.4. In particular, X is a quasimartingale on $(0,\alpha]$ or $(0,\alpha)$, with $\alpha < \infty$, if and only if the Doléans function μ_X has bounded variation $|\mu_X|$ on $\mathcal{A}(0,\alpha]$ or $\mathcal{A}(0,\alpha)$ respectively.

We say that X is of class (D) if the family $\{X_T;\ T \text{ simple stopping time}\}$ is uniformly integrable. We say X is of class (LD) if for each fixed $\alpha < \infty$, the above family, with $T \le \alpha$, is uniformly integrable.

4. Regularity of quasimartingales.

In this section we shall present one of the main theorems concerning the existence of cadlag modifications of a quasimartingale.

Recall that a family $Z \subset E^*$ is called total (for E) if $x \in E$ and $\langle x,z \rangle = 0$ for all $z \in Z$ implies that $x=0$. A family

$Z \subset E^*$ is called *norming* (for E) if for each $x \in E$ we have

$$\|x\| = \sup\{|\langle x,z\rangle|/\|z\|;\ z \in Z,\ z \neq 0\}.$$

Evidently, any norming set is total.

We remark that the RNP of E in the next theorem is used *only* in steps 7 and 8 of the proof, and is not needed in the hypotheses of the Lemmas established in the course of the proof, except for Lemma 4.II(b) and Lemma 4.VI.

4.1. *Theorem. Assume that E has the RNP and that X is a quasimartingale on each bounded interval and has a separable range. Let $Z \subset E^*$ be any set which is norming for the range of X.*

The following assertions are equivalent:

(1) *X has a cadlag modification;*

(2) *X has a right continuous modification;*

(3) *X is right continuous in the mean, that is, for the strong topology of L^1_E;*

(4) *X is right continuous for the weak topology of L^1_E;*

(5) *X is right continuous in probability;*

(6) $\lim_{t\downarrow s} E(1_F X_t) = E(1_F X_s)$ *for $s<\infty$ and $F \in \mathcal{F}_s$;*

(7) *$X = M+A$, where M is a (not necessarily right continuous) local martingale and A is a predictable, cadlag process with finite variation and $A_0 = 0$;*

(1') *$\langle X,z\rangle$ has a cadlag modification for each $z \in Z$;*

(2') *$\langle X,z\rangle$ has a right continuous modification for each $z \in Z$;*

(3') *$\langle X,z\rangle$ is right continuous in the mean, that is, in the strong topology of L^1, for each $z \in Z$;*

(4') <u>$\langle X,z\rangle$ is right continuous in the weak topology of L^1, for each $z \in Z$;</u>

(5') <u>$\langle X,z\rangle$ is right continuous in probability for each $z \in Z$;</u>

(6') $\lim_{t \downarrow s} \langle E(1_F X_t), z\rangle = \langle E(1_F X_s), z\rangle$ <u>for $s<\infty$, $F \in \mathcal{F}_s$, and $z \in Z$;</u>

(7') <u>$\langle X,z\rangle = M(z)+A(z)$ for each $z \in Z$, where $M(z)$ is a (not necessarily right continuous) local martingale and $A(z)$ is a predictable, cadlag process of finite variation with</u> $A_0(z)=0$.

<u>The decomposition in (7) or (7') is unique up to an evanescent set.</u>

<u>Proof</u>. The proof will proceed along the following implications: $7 \Rightarrow 1 \Rightarrow 2 \Rightarrow 3 \Leftrightarrow 5 \Rightarrow 6 \Rightarrow 6'$, $7 \Rightarrow 7' \Rightarrow 1' \Rightarrow 2' \Rightarrow 3' \Leftrightarrow 5' \Rightarrow 6' \Rightarrow 7$, $3 \Rightarrow 4 \Rightarrow 6'$ and $3 \Rightarrow 4' \Rightarrow 6'$. The only implications that require the space E to have the RNP are $6 \Rightarrow 7$ and $6' \Rightarrow 7$. All the other implications mentioned above are valid for any Banach space E.

The implications $7 \Rightarrow 1$ and $7' \Rightarrow 1'$ follow from the fact that any local martingale has a cadlag modification (see theorem 5.2 below). The implications $2 \Rightarrow 3 \Leftrightarrow 5$ and $2' \Rightarrow 3 \Leftrightarrow 5'$ are stated in A.4.1. We shall prove below the implication $6' \Rightarrow 7$. All the other implications are evident. The proof will be done in several steps. In each step it will be assumed that condition 6' is satisfied. We remark that the RNP is not needed in steps 1-7. In the process of proving some of the steps, we shall establish properties which are valid in more general situations and these will be stated as independent

Lemmas. In the statements of these Lemmas, no conditions will be assumed, unless otherwise specified.

We shall prove the implication 6'⟹7 assuming first that X is a quasimartingale on $(0,\infty]$; therefore μ_X has bounded variation on $\mathcal{A}(0,\infty]$. Without loss of generality we can also assume that E is separable and that Z is norming for E and countable.

<u>Step 1.</u> <u>The measure</u> μ_X <u>satisfies the following conditions:</u>

(1.1) $\lim_{t\downarrow s}|\mu_X|((s,t]\times\Omega) = 0$ <u>for</u> $s<\infty$.
<u>and</u>
(1.2) $\lim_{t\downarrow s}\mu_X((s,t]\times F)=0$ <u>for</u> $s<\infty$ <u>and</u> $F\in\mathcal{F}_s$.

In fact, hypothesis 6' is equivalent to

$$\lim_{t\downarrow s}\langle\mu_X((s,t]\times F).\ z\rangle = 0 \text{ for } s<\infty,\ F\in\mathcal{F}_s \text{ and } z\in Z.$$

Then Step 1 will be a consequence of the following lemma.

4.I. <u>Lemma.</u> <u>Let E be any Banach space, let</u> $Z\subset E^*$ <u>be a total set, and let</u> $\mu:\mathcal{A}(0,\infty]\to E$ <u>be an additive measure with bounded variation</u> $|\mu|$. <u>Then for every</u> $s<\infty$, <u>the following assertions are equivalent:</u>

(I.1) $\lim_{t\downarrow s}|\mu|((s,t]\times\Omega) = 0$;

(I.2) $\lim_{t\downarrow s}\mu((s,t]\times F) = 0$, <u>for every</u> $F\in\mathcal{F}_s$;

(I.3) $\lim_{t\downarrow s}\langle\mu((s,t]\times F),z\rangle = 0$, <u>for</u> $F\in\mathcal{F}_s$ <u>and</u> $z\in Z$.

<u>Proof.</u> Obviously (I.1)⟹(I.2)⟹(I.3). Assume now (I.3) and prove (I.2). Since μ has bounded variation

$|\mu|$, it is strongly additive on $\mathscr{A}(0,\infty]$, that is $\mu(B_n) \to 0$ for every sequence (B_n) of disjoint sets from $\mathscr{A}(0,\infty]$ (see [1] for results concerning strong additivity). As a result, for every decreasing sequence $A_n \downarrow \emptyset$ from $\mathscr{A}(0,\infty]$, the sequence $(\mu(A_n))$ is convergent in E. In particular, for $A_n = (s,t_n]\times F$, with $t_n \downarrow s$ and $F \in \mathscr{F}_s$, the limit $x = \lim_n \mu((s,t_n]\times F)$ exists in E. By assumption (I.3), we conclude that $\langle x,z\rangle = 0$ for every $z \in Z$, which proves (I.2). Next assume (I.2) and prove (I.1) (cf. also [12]). Let $t_n \downarrow s$ and let $\varepsilon > 0$. Let (R_k) be a finite partition of $(s,t_1]\times\Omega$ consisting of disjoint predictable rectangles such that

$$|\mu|((s,t_1]\times\Omega) < \Sigma_k \|\mu(R_k)\| + \varepsilon.$$

For any set $A \in \mathscr{A}(0,t_1]$ contained in $(s,t_1]\times\Omega$ we have

$$\Sigma_k \|\mu(A\cap R_k)\| \geq \Sigma_k \|\mu(R_k)\| - \Sigma_k \|\mu(A^c \cap R_k\| \geq$$
$$\geq |\mu|((0,t_1]\times\Omega) - \varepsilon - |\mu|(A^c) = |\mu|(A) - \varepsilon.$$

If we apply the above to $A = (s,t_n]\times\Omega$, we obtain

$$|\mu|((s,t_n]\times\Omega) \leq \Sigma_k \|\mu((s,t_n]\times\Omega \cap R_k)\| + \varepsilon.$$

Now let $n \to \infty$ and obtain $\lim_n |\mu|((s,t_n]\times\Omega) \leq \varepsilon$, which proves (I.1).

<u>Step 2</u>. (a) <u>For each $z \in Z$, the quasimartingale $\langle X,z\rangle$ has a cadlag modification</u>.

(b) <u>There exists a cadlag negative submartingale</u> Φ <u>such that</u> $|\mu_X| = \mu_\Phi$ <u>on</u> $\mathscr{A}(0,\infty]$.

Assertion (a) follows immediately from Lemma 4.III

below, since $\langle X,z\rangle$ satisfies condition 6' which is equivalent to condition III.1 of this lemma, for $\langle X,z\rangle$. Assertion (b) will be a corollary of the following two lemmas. In fact, from Lemma 4.II, we deduce first that $|\mu_X|$ satisfies condition (II.1), and then deduce that there exists a negative submartingale Φ such that $|\mu_X| = \mu_\Phi$ on $\mathcal{A}(0,\infty]$. By Step 1, $|\mu_X|$ satisfies condition (III.2) of Lemma 4.III, which insures that Φ can be chosen to be cadlag.

4.II. Lemma. (a) Let E be any Banach space and let Y be an adapted E-valued process with $Y_t \in L^1_E(P)$ for all $t > 0$. Then Y is a quasimartingale on $(0,\infty]$ if and only if its Doléans measure $\mu = \mu_Y$ satisfies the following condition:

(II.1) For every $t<\infty$, there is a function $f_t > 0$ in $L^1(P)$ such that $|\mu|((t,\infty]\times F) < E(1_F f_t)$, for $F \in \mathcal{F}_t$.

(b) Assume that E has RNP and let $\mu: \mathcal{A}(0,\infty] \to E$ be an additive measure with bounded variation $|\mu|$ satisfying condition (II.1) above. Then there is an E-valued quasimartingale Y such that $\mu = \mu_Y$.

Proof. (a). Assume first that Y is a quasimartingale on $(0,\infty]$. For any predictable rectangle $(u,v]\times F$, set $I_Y((u,v]\times F) = 1_F(Y_v - Y_u)$. Then I_Y is an $L^1_E(P)$-valued, finitely additive set function defined on the semi-ring of predictable rectangles; it can be extended to an additive measure on $\mathcal{A}(0,\infty]$. For every $A \in \mathcal{A}(0,\infty]$, we have $\mu_Y(A) = E(I_Y(A))$. Let $t<\infty$ and let $\mathcal{H}$

be the set of all functions in $L^1(P)$ of the form

$$\Sigma_i \| E(I_Y(A_i) | \mathcal{F}_t \|$$

where (A_i) is a finite family of disjoint predictable rectangles contained in $(t,\infty]\times\Omega$. Note that for each such family, we have

$$\Sigma_i \| I_Y(A_i) \|_1 \le |\mu_Y|((0,\infty]\times\Omega) < \infty,$$

hence the set $\mathcal{H}$ is bounded in $L^1(P)$. We next show that $\mathcal{H}$ is directed for the usual order on $L^1(P)$. If (A_i) and (B_j) are two finite families of disjoint predictable rectangles contained in $(t,\infty]\times\Omega$, and if (B_j) is a refinement of (A_i), then using the additivity of I_Y we obtain

$$\Sigma_i \| E(I_Y(A_i) | \mathcal{F}_t) \| \le \Sigma_j \| E(I_Y(B_j) | \mathcal{F}_t) \| .$$

This result then implies that $\mathcal{H}$ is directed in $L^1(P)$. $\mathcal{H}$ being bounded and directed in the space $L^1(P)$ of equivalence classes of integrable functions, it has a supremum in $L^1(P)$, and we choose a representative f_t in this equivalence class. For any set $F \in \mathcal{F}_t$ and any finite family (A_i) of disjoint predictable rectangles contained in $(t,\infty]\times F$, we have

$$\Sigma_i \| E(I_Y(A_i) | \mathcal{F}_t) \| \le 1_F f_t \text{ a.s,}$$

hence

$$\Sigma_i \| \mu_Y(A_i) \| \le E(1_F f_t);$$

consequently

$$|\mu_Y|((t,\infty]\times F) \le E(1_F f_t).$$

Conversely, if (II.1) is satisfied for $t=0$, then μ_Y has bounded variation on $(0,\infty]\times\Omega$, hence Y is a quasimartingale on $(0,\infty]$.

(b). Let μ be an E-valued additive measure on $\mathcal{A}(0,\infty]$ with bounded variation $|\mu|$ satisfying (II.1). For fixed $t < \infty$, the mapping $F \to \mu((t,\infty]\times F)$ is a countably additive measure on $\mathcal{F}_t$ with bounded variation and is absolutely continuous with respect to P. Since E has the RNP, there is a function $Y_t \in L^1_E(P)$ such that

$$\mu((t,\infty]\times F) = -E(1_F Y_t), \text{ for } F \in \mathcal{F}_t.$$

It follows that $\mu = \mu_Y$ and since μ has bounded variation on $\mathcal{A}(0,\infty]$, Y is a quasimartingale on $(0,\infty]$.

Remark. In the scalar case, a property similar to (II.1) was proved in [10], in a different setting.

In the scalar case, the following lemma characterizes the right continuity of Y in terms of the corresponding measure μ_Y. The vector case will be considered later (Theorem 4.3).

4.III. Lemma. Let μ be a real valued, additive measure on $\mathcal{A}(0,\infty]$ with bounded variation $|\mu|$ satisfying condition (II.1) and let Φ be a real valued quasimartingale on $(0,\infty]$ with $\mu = \mu_\Phi$. Then Φ has a cadlag modification if and only if

(III.1) $\quad \lim_{t\downarrow s}\mu((s,t]\times F)=0$ for $s<\infty$ and $F\in\mathcal{F}_s$.

If, moreover, $\mu > 0$ on $\mathscr{A}(0,\infty]$, then Φ is a negative submartingale; Φ has a cadlag modification if and only if

(III.2) $\quad \lim_{t\downarrow s}\mu((s,t]\times\Omega) = 0$ *for* $s<\infty$.

Proof. If Φ has a cadlag modification, then, by A.4.1, Φ is right continuous in the mean, hence $\lim_{t\downarrow s} E(1_F\Phi_t) = E(1_F\Phi_s)$ for $s<\infty$ and $F \in \mathscr{F}_s$, and this condition is equivalent to (III.1); if $\mu > 0$, this condition is equivalent to (III.2).

To prove the converse implication, assume first that $\mu > 0$ on $\mathscr{A}(0,\infty]$ and satisfies (III.2). Then Φ is a negative submartingale and (III.2) is equivalent to

$$\lim_{t\downarrow s}E(\Phi_t) = E(\Phi_s) \text{ for } s < \infty.$$

By [3,VI.4], Φ has a cadlag modification. If μ is real valued and satisfies condition (III.1), then by Lemma I, its variation $|\mu|$ satisfies (III.2). By the first part of the proof, there are two cadlag negative submartingales Φ^+ and Φ^- satisfying $\mu^+ = \mu_{\Phi^+}$ and $\mu^- = \mu_{\Phi^-}$. Then $\mu_\Phi = \mu_{(\Phi^+ - \Phi^-)}$, hence $\Phi^+ - \Phi^-$ is a cadlag modification of Φ.

Remark. A similar result was proved in [10] for measures satisfying a certain "condition(S)".

The above lemma will be extended to quasimartingales with values in a Banach space (Theorem 4.3).

For every simple stopping time T, X_T is an $\mathscr{F}_T$-measurable, integrable random variable. We shall extend, in Step 3 below, the definition of X_T by means of a random variable $\tilde{X}_T$, where T is an arbitrary stopping time T. To do this we need the following lemma.

4.IV. Lemma. If Y is an E-valued quasimartingale on $(0,\infty]$, then for every stopping time T and for every decreasing sequence (T_n) of simple stopping times with $T_n \downarrow T$, the sequence $(E(Y_{T_n} \mid \mathscr{F}_T))$ is Cauchy in L^1_E.

Proof. We remark first that if $T \leq U \leq V$, where U and V are simple stopping times, then

$$\|E((Y_V - Y_U) \mid \mathscr{F}_T)\|_1 \leq |\mu_Y|((U,V]).$$

In fact,

$$\int \|E((Y_V-Y_U)|\mathscr{F}_T)\| dP = \sup\Sigma_i \|\int_{F_i} E((Y_V-Y_U)|\mathscr{F}_T)dP\| =$$

$$= \sup\Sigma_i \|\int_{F_i} (Y_V-Y_U)dP\| = \sup\Sigma_i \|\mu_Y((U,V]\cap R_+\times F_i))\|$$

$$\leq |\mu_Y|((U,V]),$$

where the supremum is taken over all finite families (F_i) of disjoint sets from $\mathscr{F}_T$. Then, taking $U=T_{n+1}$ and $V=T_n$, we obtain

$$\Sigma_n \|E((Y_{T_n} - Y_{T_{n+1}})|\mathscr{F}_T)\|_1 \leq \Sigma_n |\mu_Y|((T_{n+1},T_n]) \leq$$
$$\leq |\mu_Y|(R_+\times\Omega) < \infty.$$

It follows that the series $\Sigma_n E((Y_{T_n} - Y_{T_{n+1}})|\mathscr{F}_T)$ is convergent in L^1_E. For any k we have

$$\Sigma_{n<k} E((Y_{T_n} - Y_{T_{n+1}})|\mathscr{F}_T) = E(Y_{T_1}|\mathscr{F}_T) - E(Y_{T_k}|\mathscr{F}_T),$$

hence $E(Y_{T_k}|\mathscr{F}_T)$ converges in L^1_E, and this proves Lemma 4.IV.

Remark. This lemma was proved in [11] under the assumption that Y is right continuous in the mean.

Step 3. For any stopping time T we can choose an $\mathcal{F}_T$-measurable random variable $\tilde{X}_T \in L^1_E$ satisfying the following conditions:

(a) *If (T_n) is a decreasing sequence of stopping times with $T_n \downarrow T$, then $E(\tilde{X}_{T_n} \mid \mathcal{F}_T) \to \tilde{X}_T$ in the mean.*

(b) *If T is a simple stopping time, then $\tilde{X}_T = X_T$ a.s.*

(c) *If X is right continuous, then $\tilde{X}_T = X_T$ a.s.*

(d) *If X' is a modification of X, then $\tilde{X}'_T = \tilde{X}_T$ a.s.*

(e) *If $z \in Z$ and $Y = \langle X, z\rangle$, then $\tilde{Y}_T = \langle \tilde{X}_T, z\rangle$ a.s.*

(f) *If $T_n \uparrow \infty$, then for any $t < \infty$ we have $\tilde{X}_{T_n \wedge t} \to X_t$ a.s.*

We first take a decreasing sequence (T_n) of simple stopping times T_n with $T_n \downarrow T$. By Lemma 4.IV, the sequence $(E(X_{T_n} \mid \mathcal{F}_T))$ is Cauchy in L^1_E, and has a limit $u \in L^1_E$. This limit is independent of the sequence (T_n). In fact, let (S_n) be another decreasing sequence of simple stopping times converging to T, and let $v = \lim_n E(X_{S_n} \mid \mathcal{F}_T)$ in L^1_E. Let $z \in Z$. By Step 2, the real quasimartingale $Y = \langle X, z\rangle$ has a cadlag modification W. By A.4.2, we have $W_T \in L^1$ and $W_{T_n} \to W_T$ and $W_{S_n} \to W_T$ in L^1; consequently both sequences $\langle E(X_{T_n} \mid \mathcal{F}_T), z\rangle$ and $\langle E(X_{S_n} \mid \mathcal{F}_T), z\rangle$ converge to W_T in L^1. Thus $\langle u, z\rangle = W_T = \langle v, z\rangle$ a.s. Since Z is countable, we have $u = v$ a.s.

We denote by $\tilde{X}_T$ a representitive of the element $\lim E(X_{T_n} \mid \mathcal{F}_T)$ in L^1_E. Then $\tilde{X}_T$ is $\mathcal{F}_T$-measurable.

For $z \in Z$ and $Y = \langle X, z\rangle$, we then have

$$\langle \tilde{X}_T, z\rangle = \langle \lim E(X_{T_n} \mid \mathcal{F}_T), z\rangle = \lim E(\langle X_{T_n}, z\rangle \mid \mathcal{F}_T) = \tilde{Y}_T,$$

and this proves (e).

If X' is a modification of X, then $X'_{T_n} = X_{T_n}$ a.s., for simple stopping times T_n and (d) follows. Also (b) follows in a similar fashion.

If X is right continuous, then (c) follows by A.4.2.

Property (a) holds by definition if (T_n) is a sequence of simple stopping times. Now let (T_n) be an arbitrary decreasing sequence of stopping times converging to T, and prove that $E(\tilde{X}_{T_n} | \mathcal{F}_T) \to \tilde{X}_T$ in L^1_E. For each n choose a simple stopping time S_n with $T_n \leq S_n \leq T_n + 1/n$ and

$$\| E(X_{S_n} | \mathcal{F}_{T_n}) - \tilde{X}_{T_n} \|_1 < 1/n.$$

Hence

$$\| E(X_{S_n} | \mathcal{F}_T) - E(\tilde{X}_{T_n} | \mathcal{F}_T) \|_1 < 1/n.$$

We can choose the S_n decreasing to T, hence $E(X_{S_n} | \mathcal{F}_T) \to \tilde{X}_T$ and then (a) follows.

To prove (f), let $z \in Z$ and let W be a cadlag modification of $Y = \langle X, z \rangle$, which exists by Step 2. Then a.s., $\langle \tilde{X}_{T_n \wedge t}, z \rangle = \tilde{Y}_{T_n \wedge t} = W_{T_n \wedge t}$ and $W_t = \langle X_t, z \rangle$. Let $N \subset \Omega$ be a P-negligible set such that for all $\omega \in N$, the above equalities are valid for a fixed t, for all integers n and all z in a countable norming set Z. For $\omega \in N$, choose n such that $T_n(\omega) > t$. Then $\langle \tilde{X}_{T_n \wedge t}(\omega), z \rangle = W_{T_n \wedge t}(\omega) = W_t(\omega) = \langle X_t(\omega), z \rangle$, hence $\tilde{X}_{T_n \wedge t}(\omega) = X_t(\omega)$, which proves (f).

Step 4. (a) For any stopping time T we have $\|\tilde{X}_T\| \leq -\Phi_T$ a.s.

(b) For any stopping times $S \leq T$ we have $\| E(\tilde{X}_T - \tilde{X}_S \| \leq E(\Phi_T - \Phi_S)$.

For $z \in Z$ and $A \in \mathcal{A}(0,\infty]$ we have

$$|\mu_{\langle X,z\rangle}(A)| = |\langle\mu_X(A),z\rangle| \leq \|z\|\mu_\Phi(A),$$

hence $|\langle X_t,z\rangle| \leq -\Phi_t$ a.s. for each $t \geq 0$. Let (T_n) be a sequence of simple stopping times, $T_n \downarrow T$. Then $|\langle X_{T_n},z\rangle| \leq -\Phi_{T_n}$ a.s. for each n, therefore

$$|E(\langle X_{T_n},z\rangle \mid \mathcal{F}_T)| \leq -E(\Phi_{T_n} \mid \mathcal{F}_T) \text{ a.s.}$$

Let $n \to \infty$ and obtain

$$|\langle\tilde{X}_T,z\rangle| \leq -\Phi_T \text{ a.s.}$$

Since Z is countable and norming, we have $\|\tilde{X}_T\| \leq -\Phi_T$ a.s.

For $S \leq T$ simple stopping times we have

$$\begin{aligned}\|E(\tilde{X}_T - \tilde{X}_S)\| &= \|\mu_X((S,T])\| \leq |\mu_X|((S,T]) \\ &= \mu_\Phi((S,T]) = E(\Phi_T - \Phi_S).\end{aligned}$$

Then (b) follows by approximating general stopping times by simple ones.

For any stopping time T, we define now the "stopped" process $\tilde{X}^T$ as follows: $(\tilde{X}^T)_t = \tilde{X}_{T\wedge t}$, for $t \geq 0$.

Step 5. For each stopping time T, the stopped process $\tilde{X}^T$ is a quasimartingale on $(0,\infty]$ satisfying the regularity condition

$$\lim_{t\downarrow s} E(1_F\tilde{X}^T_t) = E(1_F\tilde{X}^T_s), \text{ for } s<\infty,\ F \in \mathcal{F}_s.$$

If $S_1 \leq S_2 \leq \ldots \leq S_n$ is a finite family of simple stopping times, then by Step 4(b),

$$\Sigma_i \| E(\tilde{X}^T_{S_{i+1}} - \tilde{X}^T_{S_i}) \| = \Sigma_i \| E(\tilde{X}_{T \wedge S_{i+1}} - \tilde{X}_{T \wedge S_i}) \| \leq$$
$$\leq \Sigma_i E(\Phi_{T \wedge S_{i+1}} - \Phi_{T \wedge S_i}) \leq \mathrm{Var}_{\Phi}(0,\infty].$$

The last inequality is valid by A.4.3 since Φ is a cadlag quasimartingale. Taking the supremum for all families (S_i) as above, and using A.2.3, we obtain

$$\mathrm{Var}_{\tilde{X}^T}(0,\infty] \leq \mathrm{Var}_{\Phi}(0,\infty] < \infty,$$

hence $\tilde{X}^T$ is a quasimartingale on $(0,\infty]$.

Since Φ is cadlag, it is right continuous in the mean by A.4.1. Then by A.4.2, we have

$$\lim_{t \downarrow s} E(\Phi_{T \wedge t_F} - \Phi_{T \wedge s_F}) = 0,$$

therefore

$$\lim_{t \downarrow s} E(1_F \tilde{X}^T_t) = E(1_F \tilde{X}^T_s),$$

and this proves Step 5.

<u>Step 6.</u> <u>If S and T are stopping times with $S \leq T$, then $(\tilde{X}^T)^S$ is a modification of $\tilde{X}^S$.</u>

We remark first that, by Step 5, we can apply Step 3 to $\tilde{X}^T$ and define for every $t > 0$, $\widetilde{(\tilde{X}^T)}_{S \wedge t}$; for notational convenience, we shall abuse notation and we shall denote the latter simply by $(\tilde{X}^T)_{S \wedge t}$. Hence the process $(\tilde{X}^T)^S$ is defined by Step 5 and is a quasimartingale satisfying the regularity condition (R). Let $S_n \downarrow S$, S_n simple stopping times, and let $t > 0$; then
$S_n \wedge t \downarrow S \wedge t$ and $S_n \wedge T \wedge t \downarrow S \wedge t$, hence, since the S_n are

simple,

$$(\tilde{X}^T)^S_t = (\tilde{X}^T)_{S\wedge t} = \lim E((\tilde{X}^T)_{S_n \wedge t} | \mathcal{F}_{S\wedge t}) =$$

$$= \lim E(\tilde{X}_{T\wedge S_n \wedge t} | \mathcal{F}_{S\wedge t}) = \tilde{X}_{S\wedge t} = \tilde{X}^S_t \quad \text{in } L^1_E.$$

It follows that $(\tilde{X}^T)^S_t = \tilde{X}^S_t$ a.s., hence $(\tilde{X}^T)^S$ is a modification of $\tilde{X}^S$.

Next, for each n we define the stopping time

$$T_n = \inf \{t: -\Phi_t > n\}.$$

Since Φ is cadlag, we have $T_n \uparrow \infty$. Set $X^{(n)} = \tilde{X}^{T_n}$.

Step 7. $X^{(n)}$ is a quasimartingale of class (D) on $(0,\infty]$ and satisfies the regularity condition

$$\lim_{t\downarrow s} E(1_F X^{(n)}_t) = E(1_F X^{(n)}_s), \text{ for } s < \infty, F \in \mathcal{F}_s.$$

In fact, for any simple stopping time T we have by Step 4(a),

$$\| X^{(n)}_T \| \leq -\Phi_{T\wedge T_n} \leq n + \Phi_{T_n} \in L^1,$$

since Φ is cadlag (see A.4.2). This shows that $X^{(n)}$ is of class (D). The other assertion was proved in Step 5.

Step 8. $X^{(n)}$ can be decomposed uniquely (up to an evanescent set) in the form $X^{(n)} = M^{(n)} + A^{(n)}$, where $M^{(n)}$ is an E-valued martingale of class (D) and $A^{(n)}$ is an E-valued, predictable, cadlag process with integrable variation and with $A_0^{(n)}=0$.

This is a direct consequence of Step 6 and Lemma 4.VI below. In the proof of Lemma 4.VI we shall use the following Lemma:

4.V. Lemma. Let Y be an E-valued, adapted process such that $Y_t \in L_E^1(P)$ for each t. Then μ_Y is countably additive and of bounded variation on $\mathscr{A}(0,\infty]$ if and only if Y is a quasimartingale of class (D) on $\mathscr{A}(0,\infty]$ and satisfies

(V.1) $\lim_{t\downarrow s} E(1_F Y_t)=E(1_F Y_s)$ for $s<\infty$ and $F\in\mathscr{F}_s$.

Proof. Assume first that Y is a quasimartingale of class (D) satisfying condition (V.1). This condition is equivalent to

(V.2) $\lim_{t\downarrow s}\mu_Y((s,t]\times F)=0$ for $s<\infty$ and $F\in\mathscr{F}_s$.

We shall prove that μ_Y also satisfies the following condition:

(V.3) For any decreasing sequence H_n from $\mathscr{F}$, with $H_n\downarrow\emptyset$, we have $\lim_n \sup_{\Delta_n}\{\|\mu_Y((S,T])\|\} = 0$, where Δ_n is the family of stochastic intervals (S,T], such that S and T are stopping times, $S<T$ and $(S,T]\subset(0,\infty]\times H_n$.

In fact, for (H_n) and S,T as above, we have

$$\|\mu_Y((S,T])\| = \|E(1_{H_n}(Y_T - Y_S)\| \leq$$

$$\leq E(1_{H_n}(\|Y_T\| + \|Y_S\|),$$

and this last term tends to zero uniformly relative to

(S,T]. We now use the criteria in [12, Lemma 2, p.83] and deduce that conditions (V.2) and (V.3) imply that μ_Y is countably additive on $\mathcal{A}(0,\infty]$.

Conversely, assume that μ_Y is countably additive and of bounded variation on $\mathcal{A}(0,\infty]$. Then it is immediate that Y is a quasimartingale on $(0,\infty]$ satisfying (V.1). Since $|\mu_Y|$ is also countably additive, we have

$$\lim_{t \downarrow s} |\mu_Y|((s,t] \times \Omega] = 0 \text{ for } s < \infty.$$

By Lemma 4.III, there is a cadlag, negative submartingale Φ such that $|\mu_Y| = \mu_\Phi$. Then

$$\|E(1_F Y_t)\| \leq - E(1_F \Phi_t) \text{ for } t<\infty \text{ and } F \in \mathcal{F}_t.$$

From this, we deduce that $\|Y_t\| \leq - \Phi_t$ a.s. As a result, $\|Y_T\| \leq -\Phi_T$ a.s. for every simple stopping time T. By [12, Thm. 13.3], Φ is of class (D), hence Y is also of class (D).

Remark. Métivier [12] proved the above lemma under the additional assumption that Y is right continuous. His result follows from the above lemma since right continuity implies right continuity in the mean (see A.4.1.), which, in turn, implies (V.1).

4.VI Lemma. Assume E has the RNP. The following two assertions are equivalent:

(a) Y is a quasimartingale of class (D) on $(0,\infty]$ and satisfies

$$\lim_{t \downarrow s} E(1_F Y_t) = E(1_F Y_s), \text{ for } s < \infty \text{ and } F \in \mathcal{F}_s.$$

(b) Y = M + A, where M is an E-valued (not

<u>necessarily cadlag) martingale of class</u> (D) <u>and</u> A <u>is an</u> E-<u>valued cadlag process with integrable variation and with</u> $A_0 = 0$.

<u>The decomposition in</u> (b) <u>is unique (up to an evanescent set).</u>

<u>Proof</u>. Assume first (a). By Lemma 4.V, μ_Y is σ-additive and has bounded variation on $\mathscr{A}(0,\infty]$; therefore it can be extended to a σ-additive measure, still denoted by μ_Y on the σ-algebra of all predictable subsets of $(0,\infty]\times\Omega$. Consider the measure λ defined on the σ-algebra $\mathcal{P}$ of all predictable subsets of $(0,\infty)\times\Omega$, equal to μ_Y on the predictable subsets of $(0,\infty)\times\Omega$, and equal to 0 on the predictable subsets of $\{0\}\times\Omega$. Then λ is σ-additive and has finite variation on $\mathcal{P}$. Since E has the RNP, there is a predictable E-valued process such that $\|g\| \equiv 1$ and

$$\lambda(B) = \int_B g d|\lambda|, \text{ for } B \in \mathcal{P}.$$

Since μ_Y and λ vanish on evanescent sets, there is a unique (up to an evanescent set), predictable, cadlag, increasing, integrable process V associated to $|\lambda|$ ([3,VI.65]) by

$$|\lambda|(B) = E(\int 1_B dV), \text{ for } B \in \mathcal{P}.$$

See also [5] for details. Since $|\lambda|(\{0\}\times\Omega)=0$ we have $V_0 = 0$. Then

$$\lambda(B) = E(\int 1_B g_s dV_s) \text{ for } B \in \mathcal{P}.$$

The process $A_t = \int_{[0,t]} g_s dV_s$ is cadlag, predictable, with

$A_0 = 0$ and

$$\lambda(B) = E(\int 1_B dA) = \mu_A(B), \text{ for } B \in \mathcal{P}.$$

The variation $|A|$ of A is integrable, since it is predictable and satisfies $|A| \leq V$. For $B \in \mathcal{A}(0,\infty)$ we have

$$\mu_Y(B) = \lambda(B) = \mu_A(B).$$

If we set $M = Y - A$, then $\mu_M = 0$ on $\mathcal{A}(0,\infty)$, hence M is a martingale. Moreover, M is of class (D) since both Y and A are of class (D), and this proves (b).

The uniqueness of the decomposition follows from the fact that, (as in the scalar case), an E-valued, predictable, cadlag martingale with finite variation and vanishing at 0 is equal to 0 everywhere (except on an evanescent set).

Conversely, if we assume (b), then Y is of class (D) and $\mu_Y = \mu_A$ on $\mathcal{A}(0,\infty)$ and $\mu_Y = \mu_M$ on $\{\infty\} \times \mathcal{F}$. Since μ_M and μ_A are σ-additive and of bounded variation on $\{\infty\} \times \mathcal{F}$ and $\mathcal{A}(0,\infty)$ respectively, it follows that μ_Y is σ-additive and with bounded variation on $\mathcal{A}(0,\infty]$; hence, by Lemma 4.V, Y satisfies condition (a).

<u>Step 9</u>. <u>We have</u> X=M+A <u>as in assertion</u> (7) <u>of the statement of the theorem</u>.

By step 8, for each n we have a unique decomposition $X^{(n)} = M^{(n)} + A^{(n)}$, where $M^{(n)}$ is a martingale of class (D) and $A^{(n)}$ is a cadlag, predictable process with integrable variation and $A_0^{(n)} = 0$. Let $N^{(n)}$ be a cadlag modification of $M^{(n)}$ (see [2]). Then $Y^{(n)} = N^{(n)} + A^{(n)}$ is a cadlag modification of $X^{(n)}$. By step 6, $(X^{(n+1)})^{T_n}$ is a modification of $X^{(n)}$; $(X^{(n+1)})^{T_n}$ therefore $(Y^{(n+1)})^{T_n}$ is a modification of $Y^{(n)}$; since they are cadlag, we have $(Y^{(n+1)})^{T_n} = Y^{(n)}$ (outside an evanescent set). It follows that $(N^{(n+1)})^{T_n} = N^{(n)}$ and $(A^{(n+1)})^{T_n} = A^{(n)}$. Then we can define the cadlag processes Y, N, A such that these processes, stopped at T_n are equal, respectively with $Y^{(n)}$, $N^{(n)}$, $A^{(n)}$. It follows that N is a cadlag local martingale and A is a cadlag, predictable process with finite variation and with $A_0 = 0$, and $Y=N+A$. We deduce, using step 3(f), that Y is a modification of X. Hence the process $M = X-A$ is a modification of the local martingale N, and for each n and t we have

$$(M^{T_n})_t = X_t^{(n)} - A_t^{(n)} = N_t^{(n)} \quad \text{a.s.},$$

hence M^{T_n} is a martinglae of class (D). It follows that M is a local martingale and this proves the implication $6' \Longrightarrow 7$ in case X is a quasimartingale on $(0,\infty]$.

Step 10. Assume now that X is a quasimartingale on each bounded interval. Then for each $\alpha < \infty$, the stopped process X^α is a quasimartingale on $(0,\infty]$ satisfying assertion (6') of the theorem. By step 9, we have $X^\alpha = M^\alpha + A^\alpha$ with M^α and A^α as in assertion (7) of the theorem. By the uniqueness of this decomposition, taking $\alpha = 1,2,\ldots$, we have $(X^{n+1})^n = X^n$ hence $(M^{n+1})^n = M^n$ and $(A^{n+1})^n = A^n$; hence there is a local martingale M and a predictable cadlag process A with finite variation and $A_0=0$ such that M and A stopped at n are equal respectively with N^n and A^n; hence $X=M+A$, and the theorem is completely proved.

The next theorem gives equivalent conditions for X to be actually cadlag (rather than having a cadlag modification). In particular, it states that X is strongly cadlag if and only if it is weakly cadlag.

4.2 Theorem. Assume that E has the RNP and that X has separable range and is a quasimartingale on each bounded interval. Let $Z \subset E^*$ be a set which is norming for the range of X. The following assertions are equivalent (up to evanescent sets):

(a) X is cadlag;

(b) X is right continuous;

(c) X is weakly right continuous;

(d) $\langle X,z\rangle$ is right continuous for every $z \in Z$;

(e) $X = M+A$, where M is a cadlag local martingale and A is a cadlag predictable process of finite variation with $A_0=0$.

Proof. The implications (e)$\Rightarrow$(a)$\Rightarrow$(b)$\Rightarrow$(c)$\Rightarrow$(d) are obvious and valid in any Banach space (not necessarily having the RNP). Now assume (d) and show (e) holds. We can assume Z is countable, since the range of X is separable. By A.4.1, $\langle X,z\rangle$ is right continuous in the mean for each $z \in Z$, hence it satisfies condition (6') of Theorem 4.1:

$$\lim_{t\downarrow s} E(\langle 1_F X_t, z\rangle) = E(\langle 1_F X_s, z\rangle), \text{ for } s<\infty,\ F\in \mathcal{F}_s, z \in Z.$$

Then by Theorem 4.1, $X=N+A$, where N is a local martingale and A is a cadlag predictable process with finite variation and $A_0=0$. Let M be a cadlag modification of N (see Theorem 5.2). Then $Y = M+A$ is a cadlag modification of X. This implies that $\langle Y,z\rangle = \langle X,z\rangle$ up to an evanescent set, for each fixed $z \in Z$. Since Z is countable, $X=Y$ up to an evanescent set and the theorem follows.

Lemma 4.III can now be completed as follows:

4.3 Theorem *Assume that E has the RNP. Let $\mu: \mathscr{A}(0,\infty] \to E$ be an additive measure with bounded variation $|\mu|$. Then there exists a cadlag quasimartingale Y on $(0,\infty]$ such that $\mu = \mu_Y$ if and only if the following two conditions hold:*

(4.3.1) *For every $t > 0$, there exists a function $f_t \in L^1(P)$ such that $|\mu|((t,\infty]\times F) \leq E(1_F f_t)$, for $F \in \mathscr{F}_t$;*

and

(4.3.2) $\lim_{t\downarrow s}\mu((s,t]\times F)=0$ *for* $s<\infty$ *and* $F \in \mathscr{F}_s$.

Proof. Since E has the RNP, by Lemma 4.II, μ satisfies (4.3.1) if and only if there exists an E-valued quasimartingale W such that $\mu = \mu_W$. Then by Theorem 4.1, W has a cadlag modification Y if and only if condition (6) of Theorem 4.1 holds, which is equivalent to condition (4.3.2) above.

5. Cadlag modifications without RNP.

In this section we shall investigate the existence of right continuous or cadlag modifications of X without assuming that E has the RNP. The characterization of cadlag modifications of X is given in terms of its right limits X_+ along the rationals (any countable set dense in R_+ can be used). The following results are proved:

1.) If X is a local martingale then X_+ exists (outside an evanescent set) and is a cadlag modification of X.

2.) If X is a quasimartingale and if X_+ exists (outside an evanescent set), then X_+ is a modification of X if and only if the regularity condition (R) holds. If X is a real-valued quasimartingale, then X_+ and X_- exist (outside an evanescent set).

3.) If X is a process with integrable variation, then X_+ (which exists everywhere) is a modification of X if and only if condition (R) holds.

5a.) Right and left limits.

For $\omega \in \Omega$ and $t \in R_+$, we denote (in this section only)

$$X_{t+}(\omega) = \lim_{r\downarrow t} X_r(\omega), \quad X_{t-}(\omega) = \lim_{r\uparrow t} X_r(\omega),$$

for r rational, whenever these limits exist. We denote by X_+ and X_- the functions $(t,\omega) \to X_{t+}(\omega)$ and $(t,\omega) \to X_{t-}(\omega)$ respectively.

We notice that if X_+ is defined on a set $M = R_+ \times F$ with $F \subset \Omega$, then X_+ is right continuous on M. Moreover, X_+

is cadlag if and only if X_- also exists on M.

We know that if X is a real-valued supermartingale, then X_+ and X_- exist a.s. (i.e. outside an evanescent set) (see [3], VI.3). It follows that if X is a real-valued quasimartingale, then X_+ and X_- exist a.s. But if X is an E-valued quasimartingale, we do not know whether X_+ or X_- exist. We list first some general properties of X_+.

5.1. *Lemma.* (a) *If* X *has a right continuous (resp. cadlag) modification, then* X_+ *(resp.* X_+ *and* X_-*) exist a.s. and* X_+ *is a right continuous (resp. cadlag) modification of* X.

(b) *Let* (T_n) *be an increasing sequence of stopping times with* $T_n \uparrow \infty$, *such that* X^{T_n} *has a right continuous (resp, cadlag) modification for each* n. *Then* X_+ *(resp.* X_+ *and* X_-*) exists a.s. and* X_+ *is a modification of* X.

(c) *Let* $Z \subset E^*$ *be a countable set, norming for the range of* X. *If* X_+ *exists a.s. and if* $\langle X,z\rangle$ *has a right continuous modification for each* $z \in Z$, *then* X_+ *is a right continuus modification of* X.

Proof. (a) If Y is a modification of X, the assertion follows from the equality $X_r=Y_r$ a.s. for each r rational.

(b) For each n, $(X^{T_n})_+$ exists, by (a), and is a modification of X^{T_n}. Then X_+ exists a.s. and $(X^{T_n})_{t+}= X_t^{T_n}$ a.s. for each t; letting $n \to \infty$ we get $X_{t+} = X_t$ a.s. for each t.

(c) For each $z \in Z$, $\langle X,z\rangle_+$ exists a.s. and is a modification of $\langle X,z\rangle$. Then

$$\langle X_{t+}, z\rangle = \langle X_t, z\rangle_+ = \langle X_t, z\rangle \text{ a.s.}$$

for each $z \in Z$ and $t>0$. Since Z is countable, $X_{t+} = X_t$ a.s. for each t.

5b.) Local martingales

We call X a local martingale if there is an increasing sequence (T_n) of stopping times with $T_n \uparrow \infty$ a.s., such that for each n, X^{T_n} is a martingale (not necessarily right continuous). This means that for each n and t, $X_t^{T_n}$ is integrable; but $X_T^{T_n}$ is not necessarily integrable if T is an arbitrary stopping time. We note that a modification of a local martingale is not necessarily a local martingale.

5.2. Theorem. If X is a local martingale then X_+ and X_- exist a.s.; X_+ is a local martingale and a cadlag modification of X.

Proof. Let $T_n \uparrow \infty$ such that X^{T_n} is a martingale for each n. By [2], each X^{T_n} has a cadlag modification. Then we apply Lemma 5.1 (a) and (b) to deduce that $(X^{T_n})_+$ is a modification of X^{T_n} and X_+ is a modification of X. Moreover, X_+ is a local martingale, since for each n we have $(X_+)^{T_n} = (X^{T_n})_+$ and $(X^{T_n})_+$ is a martingale.

5c.) Quasimartingales

We first state a theorem on real-valued quasimartingales, which seems to be new.

5.3. *Theorem.* *If* X *is a real valued quasimartingale on bounded intervals, then* X_+ *and* X_- *exist a.s. and are quasimartingales.* X_+ *is a cadlag modification of* X *if and only if condition* (R) *holds:*

(R) $\lim_{t\downarrow s} E(1_F X_t) = E(1_F X_s)$ *for* $s<\infty$ *and* $F \in \mathcal{F}_s$.

Proof. Without loss of generality we can assume that X is a quasimartingale on $(0,\infty]$. Then μ_X has bounded variation $|\mu_X|$ on $\mathcal{A}(0,\infty]$ and μ_X^+ and μ_X^- are positive additive measures on $\mathcal{A}(0,\infty]$. Let Φ and Ψ be negative submartingales such that $\mu_X^+ = \mu_\Phi$ and $\mu_X^- = \mu_\Psi$. Then Φ_+, Φ_-, Ψ_+ and Ψ_- exist a.s. and are submartingales ([3], VI.3). From the equality $\mu_X = \mu_{\Phi-\Psi}$ on $\mathcal{A}(0,\infty]$, we deduce that $\Phi-\Psi$ is a modification of X. It follows that X_+ and X_- exist a.s. and are modifications of the quasimartingales $\Phi_+ - \Psi_+$ and $\Phi_- - \Psi_-$ respectively; hence X_+ and X_- are also quasimartingales.

If X_+ is a right continuous modification of X, then X is right continuous in the mean (A.4.1) and condition (R) follows.

Conversely, if condition (R) holds, then, by Lemma 4.III, X has a cadlag modification; by Lemma 5.1(a), X_+ is a cadlag modification of X.

5.4. *Theorem.* *Assume that* X *is a quasimartingale on bounded sets, has separable range and* X_+ *exists a.s.* *Then* X_+ *is a right continuous modification of* X *if and only if condition* (R) *holds.*

Proof. Assume first that X satisfies condition (R)

and let $Z \subset E^*$ be a countable set, norming for the range of X. Then for each $z \in Z$, $\langle X,z\rangle$ is a real-valued quasimartingale satisfying condition (R). By theorem 5.3, $\langle X,z\rangle_+$ exists a.s. and is a cadlag modification of $\langle X,z\rangle$. Then, by Lemma 5.1(a), X_+ is a right continuous modification of X.

Conversely, if X_+ is a right continuous modification of X, then X is right continuous in the mean (A.4.1) and condition (R) follows.

5d.) Processes with finite variation.

The following theorem is new even in the scalar case. We remark that a modification of a process with finite variation does not necessarily have finite variation.

5.5. Theorem. Assume X has finite variation $|X|$ and that $|X|_t$ is integrable for every t. Then X_+ (which exists everywhere) has integrable variation on bounded intervals. Moreover, X_+ is a modification of X if and only if condition (R) holds.

Proof. The existence of X_+ and X_- follows from the inequality $\|X_t - X_s\| \leq |X|_t - |X|_s$ for $s < t$. From this inequality we deduce $\|X_{t+} - X_{s+}\| \leq |X|_{t+} - |X|_{s+}$ for $s<t$. Since $|X|_t$ is increasing, it follows that X_+ has finite variation $|X_+|$, and that $|X_+| \leq |X|_+$. X_+ is right continuous and adapted, hence its variation $|X_+|$ is right continuous and adapted. From $|X_+|_t \leq |X|_{t+} \leq |X|_{t+1} \in L^1$ we

deduce that $|X_+|_t \in L^1$ for every $t>0$. Now, since X has integrable variation on bounded intervals, it is a quasimartingale on bounded intervals. By theorem 5.4, X_+ is a cadlag modification of X if and only if condition (R) holds, and the theorem is proved.

Remark. If the variation $|X|_t$ of X on $[0,t]$ is integrable, then it is $\mathscr{F}_t$-measurable. In fact, the set Φ of functions of the form $\Sigma_i \|X_{t_{i+1}} - X_{t_i}\|$ with $0=t_0<\ldots<t_n=t$ is directed and bounded in $L^1(\mathscr{F}_t)$, hence $\sup\Phi$ exists in $L^1(\mathscr{F}_t)$. Since $|X|_t = \sup\Phi$ a.s. it follows that $|X|_t$ is $\mathscr{F}_t$-measurable.

6) The Doob-Meyer decomposition of quasimartingales

Theorems 4.1 and 4.2 can be viewed as giving equivalent conditions for the Doob-Meyer decomposition $X=M+A$ of a quasimartingale, in terms of right continuity. Lemma 4.VI gives conditions for the Doob-Meyer decompostition of quasimartingales of class (D). The following theorem is an extension of both Lemmas 4.V and 4.VI for quasimartingales of class (LD).

6.1 Theorem. The following two assertions are equivalent:

(a) X is a quasimartingale on bounded intervals, is of class (LD) and satisfies

(R) $\lim_{t\downarrow s} E(1_F X_t) = E(1_F X_s)$ for $s<\infty$ and $F \in \mathscr{F}_s$;

(b) μ_X has bounded variation on each ring $\mathscr{A}(0,\alpha]$ with $\alpha<\infty$, and is σ-additive on $\mathscr{A}(0,\infty)$.

If E *has the* RNP, *the above assertions are equivalent to*

(c) X=M+A, *where* M *is a (not necessarily right continuous or bounded) martingale and* A *is a cadlag, predictable process with* $A_0=0$ *and with integrable variation on each bounded interval. The decomposition is unique up to an evanescent set.*

For the correspondence $X \leftrightarrow \mu_X \leftrightarrow M+A$ established above we can state the following completions:

6.2. X *is a quasimartingale on* $(0,\infty)$ *of Class* (LD) *and satisfies condition* (R) *iff* μ_X *is* σ-*additive and of bounded variation on* $\mathcal{A}(0,\infty)$ *iff* M *is a martingale and* A *has integrable variation.*

6.3. X *is a quasimartingale on* $(0,\infty]$ *of class* (LD) *and satisfies condition (R) iff* μ_X *has bounded variation on* $\mathcal{A}(0,\infty]$ *and is* σ-*additive on* $\mathcal{A}(0,\infty)$ *iff* M *is a martingale bounded in* L^1_E *and* A *has integrable variation.*

6.4. X *is a quasimartingale on* $(0,\infty]$ *of class* (D) *and satisfies condition* (R) *iff* μ_X *has bounded variation and is* σ-*additive on* $\mathcal{A}(0,\infty]$ *iff* M *is a martingale of class* (D) *and* A *has integrable variation.*

Proof. Statement 6.4 is proven in Lemmas 4.V and 4.VI. To prove theorem 6.1, we apply 6.4 for each $n \in N$ to the stopped quasimartingale X^n. The implication a$\Rightarrow$b follows then by noticing that if μ_X is σ-additive on each ring $\mathscr{A}(0,n]$, then it is σ-additive on $\mathscr{A}(0,\infty)$. The implication a$\Rightarrow$c is proved by decomposing $X^n=M^n+A^n$ where M^n is a martingale of class (D) and A^n is a predictable, cadlag process with integrable variation and $A_0^n=0$. From the uniqueness of this decomposition we deduce the existence of two processes M and A which stopped at n are equal to M^n and A^n respectively. Finally we remark that M is itself a martingale. The implications b$\Rightarrow$a and c$\Rightarrow$a are evident. Statements 6.2 and 6.3 are immediate.

APPENDIX

In this section we shall collect some properties of the concepts involved in this paper.

A.1. **The Doléans function.**

The following properties hold for the Doléans function μ_X introduced in section 2.

A.1.1. X is a martingale if and only if $\mu_X=0$ on $\mathscr{A}(0,\infty)$.

A.1.2. X is a submartingale if and only if $\mu_X > 0$ on $\mathscr{A}(0,\infty)$. X is a negative submartingale if and only if $\mu_X > 0$ on $\mathscr{A}(0,\infty]$.

A.1.3. If Y is another process like X, then $\mu_X = \mu_Y$ on $\mathscr{A}(0,\infty)$ if and only if X-Y is a martingale. We have $\mu_X = \mu_Y$ on $\mathscr{A}(0,\infty]$ if and only if X is a modification of Y.

A.1.4. If Y is a negative submartingale and if $\|\mu_X(A)\| \leq \mu_Y(A)$, for $A \in \mathscr{A}(0,\infty]$, then $\|X_t\| \leq -Y_t$ a.s. for each t.

A.1.5. For any stochastic interval (S,T], with S and T simple stopping times, we have $\mu_X(S,T] = E(X_T - X_S)$.

A.1.6. Assume X is a quasimartingale on $(0,\alpha]$, for each $\alpha<\infty$ and let Z be a total subset of E^*. The following conditions are equivalent for any $s<\infty$.

(a) $\lim_{t\downarrow s} |\mu_X|((s,t]\times\Omega) = 0$;

(b) $\lim_{t\downarrow s} \mu_X((s,t]\times F) = 0$ for any $F \in \mathscr{F}_s$;

(c) $\lim_{t\downarrow s} \langle \mu_X((s,t]\times F), z\rangle = 0$, for $F \in \mathscr{F}_s$ and $z \in Z$.

(d) $\lim_{t\downarrow s} E(1_F X_t) = E(1_F X_s)$, for $F \in \mathscr{F}_s$;

(e) $\lim_{t\downarrow s} \langle E(1_F X_t), z\rangle = \langle E(1_F X_s), z\rangle$, for $F \in \mathscr{F}_s$ and $z \in Z$.

The equivalence of (a), (b), (c) was proved in Lemma I. The equivalences (b)$\Longleftrightarrow$(d) and (c)$\Longleftrightarrow$(e) are evident and are valid even if X is not a quasimartingale.

A.2 <u>Mean variation.</u>

The mean variation was introduced in section 3.

A.2.1. Proposition. For any $\alpha < \infty$, we have

$$\mathrm{Var}_X(0,\alpha] = \sup \Sigma \| E(1_{F_i}(X_{t_i} - X_{s_i}) | \mathscr{F}_{s_i}) \|_1,$$

where the supremum is taken over all finite families $((s_i,t_i]\times F_i)$ of disjoint predictable rectangles contained in $(0,\alpha]\times\Omega$.

To prove the above, arrange the endpoints s_i and t_i in an increasing order, $0 \leq u_1 \leq \ldots \leq u_n = \alpha$. Then

$$(s_i,t_i]\times F_i = \bigcup_{1\leq j\leq n}(u_j,u_{j+1}]\times A_{ij},$$

where $A_{ij} = \emptyset$ if $(s_i,t_i]\cap(u_j,u_{j+1}] = \emptyset$ and $A_{ij} = F_i \in \mathscr{F}_{u_j}$ if $s_i \leq u_j \leq t_i$. For fixed j, the sets A_{ij} and A_{kj} are disjoint if $i \neq k$. Then

$$\Sigma_i \| E(1_{F_i}(X_{t_i} - X_{s_i}) | \mathscr{F}_{s_i}) \|_1 \leq \Sigma_j \| E(X_{u_{j+1}} - X_{u_j} | \mathscr{F}_{u_j}) \|_1 \leq$$
$$\leq \mathrm{Var}_X(0,\alpha],$$

and this inequality is preserved if we take the supremum in the left hand side. The converse inequality is evident.

A.2.2. Proposition. For any $\alpha < \infty$, we have

$$|\mu_X|((0,\alpha]\times\Omega) = \mathrm{Var}_X(0,\alpha]$$

and

$$|\mu_X|((0,\alpha)\times\Omega) = \mathrm{Var}_X(0,\alpha).$$

For the proof, use A.2.1 above and theorem 6, page 232 in [4]:

$$\| E(X_{t_{i+1}} - X_{t_i} | \mathscr{F}_{t_i}) \|_1 = \sup | E(\langle E(X_{t_{i+1}} - X_{t_i} | \mathscr{F}_{t_i}), f\rangle) |$$

the sup being over the $\mathcal{F}_{t_i}$-measurable step functions f with $\|f\|_\infty \leq 1$. See [16,2D1] for the real case and [12,ex. E11.2].

A.2.3 Proposition. For any $\alpha \leq \infty$, we have

$$\mathrm{Var}_X(0,\alpha] = \sup \Sigma_i \|E(X_{S_{i+1}} - X_{S_i})\| = $$
$$= \sup \Sigma_i \|E(X_{S_{i+1}} - X_{S_i} | \mathcal{F}_{S_i})\|_1,$$

the supremum being taken over all finite increasing sequences $S_1 \leq \ldots \leq S_n$ of simple stopping times S_i, with $S_n \leq \alpha$.

The proof is similar to that of the scalar case [10,Th.9.2]. If X is a right continuous quasimartingale, the above equalities remain valid for arbitrary stopping times S_i (see A.4.3).

A.3. Quasimartingales

For the definition of quasimartingales see section 3.

A.3.1. X is a quasimartingale on $(0,\alpha)$ or on $(0,\alpha]$ if and only if μ_X has bounded variation on $\mathcal{A}(0,\alpha)$ or $\mathcal{A}(0,\alpha]$ respectively.

A.3.2. Any martingale is a quasimartingale on $(0,\infty)$; it is a quasimartingale on $(0,\infty]$ if and only if $\sup_t \|X_t\|_1 < \infty$; in this case $\mathrm{Var}_X(0,\infty] = \sup_t \|X_t\|_1$.

A.3.3. Any negative submartingale and any positive supermartingale is a quasimartingale on $(0,\infty]$.

A.3.4. A process X with integrable variation (that is, $E(|X|_{\infty-})<\infty$) is a quasimartingale on $(0,\infty]$.

A.3.5. X is a quasimartingale on $(0,\infty]$ if and only if X is a quasimartingale on $(0,\infty)$ and $\sup_t \|X_t\|_1<\infty$.

A.3.6. X is a quasimartingale on $(0,\alpha]$ if and only if the stopped process X^α is a quasimartingale on $(0,\infty]$.

A.3.7. If X is a quasimartingale on $(0,\alpha]$, then $\|X\|$ is a quasimartingale on $(0,\alpha]$.

A.3.8. Proposition. X is a quasimartingale on $(0,\infty)$ (respectively on $(0,\infty]$) if and only if the following condition is satisfied:

for every $t<\infty$, there is a positive function $f_t \in L^1(P)$ such that for every $F\in \mathscr{F}_t$ we have

$$|\mu_X|((t,\infty)\times F) \leq E(1_F f_t) \quad (\text{resp. } |\mu_X|((t,\infty]\times F \leq E(1_F f_t)).$$

This is lemma 4.II for quasimartingales on $(0,\infty]$.

A.4. **Right continuous quasimartingales.**

A.4.1. Proposition. Assume X is a quasimartingale on each bounded interval. X is right continuous in the mean if and only if it is right continuous in probability. If X has a right continuous modification, then it is right continuous in the mean.

The above proposition is due to Orey [15] for the scalar case. The proof for the vector case is similar.

See also [10,Th.9.3].

A.4.2. Proposition. Assume X is a right continuous quasimartingale on $(0,\infty]$. Then

(a) X_T is integrable for every stopping time T;

(b) If T_n is a decreasing sequence of stopping times converging to T, then $X_{T_n} \to X_T$ in the mean.

This is proved by Orey [15] in the scalar case. See also [10,Th.9.4].

A.4.3. Proposition. Assume X is a right continuous quasimartingale. Then the equalities in Proposition A.2.3 remain valid for arbitrary stopping times S_i, with $S_i \leq \alpha$.

This follows from Propositions A.2.3 and A.4.2. See also [10,Th.9.2] for the scalar case.

A.4.4. Proposition. Assume X is a quasimartingale on each bounded interval, and let $Z \subset E^*$ be a total set for E. We have

(R) $\lim_{t\downarrow s} E(1_F X_t) = E(1_F X_s)$, for each $s<\infty$ and $F \in \mathcal{F}_s$,

in case any one of the following conditions is satisfied for every $z \in Z$:

(1) $\langle X,z\rangle$ (in particular X) has a right continuous modification;

(2) $\langle X,z\rangle$ (in particular X) is right continuous in the mean;

(3) $\langle X,z\rangle$ (in particular X) is right continuous in

probability;

(4) $\langle\mu_X, z\rangle$ (*in particular* μ_X) *is countably additive.*

Indication of proof. By A.4.1, (1)$\Rightarrow$(2) and (2)$\Leftrightarrow$(3). Then (2) implies, for every $z \in Z$, $s<\infty$ and $F \in \mathcal{F}_s$

$$\lim_{t\downarrow s} E(1_F\langle X_t, z\rangle) = E(1_F\langle X_s, z\rangle)$$

which is equivalent to

$$\lim_{t\downarrow s}\langle\mu_X((s,t]\times F, z\rangle = 0.$$

Then apply Lemma 4.I to deduce (R). Finally, (4)$\Rightarrow$(R) is evident.

Bibliography

1. J.K. Brooks and N. Dinculeanu, Strong additivity, absolute continuity and compactness in spaces of measures, J. Math. Analysis and Appl. 45(1974), 156-175.

2. ----------- Projections and regularity of abstract processes, Stochastic Analysis and Applications, 5(1) (1987), 17-25.

3. C. Déllacherie and P. Meyer, Probabilities and Potential, North Holland, 1978, 1982.

4. N. Dinculeanu, Vector measures, Pergamon Press, 1967.

5. ----------- Vector valued stochastic processes I. Vector measures and vector-valued processes with finite variation, Journal of Theoretical Probability, 1(1988).

6. D.L. Fisk, Quasimartingales, Trans. AMS, 120(1965), 359-388.

7. H. Fölmer, The exit measure of a supermartingale, Z.W. 21(1972), 154-166.

8. T. Jeulin, Partie positive d'une quasimartingale, CRAS, Paris, 287(1978), 351.

9. H. Kunita, Stochastic integrals based on martingales taking values in Hilbert spaces, Nagoya Math. J., 38(1970), 41-52.

10. A.U. Kussmaul, Stochastic integration and generalized martingales, Pitman, London, 1977.

11. ----------- Regularität und Stochastische Integration von Semimartingalen mit Werten in einen Banachraum, Dissertation, Stuttgart, 1978.

12. M. Métivier, Semi-martingales, Walter de Gruyter, Berlin, 1982.

13. M. Métivier et J. Pellaumail, On Doléans-Fölmer's measure for quasimartingales, Illinois J. Math. 77(1975), 491-504.

14. ----------- Stochastic Integration, Academic Press, New York, 1980.

15. S. Orey, F-processes, Proc. 5th Berkley Symposion 2, (1965-66), 301-313.

16. J. Pellaumail, Sur l'intégration stochastique et la décomposition de Doob-Meyer, Astérisque 9, (1973), Société Math. de France.

17. K.M. Rao, Quasimartingales, Math. Scand. 24(1969), 79-92.

18. C. Stricker, Quasimartingales, martingales locales, semimartingales et filtration naturelles, Z.W. 39(1977), 55-64.

19. Ch. Yoeurp, Compléments sur les temps locaux et les quasimartingales, Astérisque 52-53 (1978).

20. M. Yor, Quelques interactions entre mesures vectorielles et intégrales stochastiques, Sem. Th. du Potentiel, 4(1979), Springer, Lecture Notes 713, 264-281.

J.K. Brooks and N. Dinculeanu

Department of Mathematics

University of Florida

Gainesville, Florida 32611

AUTOUR DES ENSEMBLES SEMI-POLAIRES

par

C. DELLACHERIE

Mon exposé de Princeton se composait de deux parties : une nouvelle démonstration des théorèmes à la Hunt sur les noyaux vérifiant le principe complet du maximum (en abrégé, le PCM), rédigée en détail dans [7], et un procédé de construction de noyaux vérifiant le PCM mais trop explosifs pour être des noyaux potentiels, qui n'est qu'esquissée dans [7]. Je vais ici développer cette seconde partie, et comme elle repose entr'autres sur le résultat fin suivant concernant les ensembles semi-polaires

> Si un semi-groupe de Hunt admet une probabilité de référence λ diffuse, alors il existe une probabilité μ orthogonale à λ, ne chargeant pas les semi-polaires, et chargeant tous les ouverts fins non vides.

je vais en profiter pour rassembler en un même lieu tout ce que j'ai appris en vingt ans sur ces ensembles, et que j'ai plus ou moins publié au fil des années.

En 1979, à Standford, j'avais, sur l'invitation de Chung, fait et rédigé une conférence sur les ensembles semi-polaires, que, malgré une diffusion "clandestine", je ne pus me résoudre à publier pour diverses raisons, la principale étant qu'il me restait quelques problèmes à élucider quand l'hypothèse (L) n'était pas vérifiée. Je ne vais pas tenter aujourd'hui de surmonter ces vieilles difficultés : je vais simplement reprendre mon manuscrit de 1979, en l'actualisant légèrement [je ne parlerai pas des travaux de Hansen [17], ni de ceux de Ancona, Mokobodzki et surtout Feyel [14], en théorie "pure" du potentiel, vers 1981], et en le complétant par ce qui a été annoncé plus haut. Pour rester dans des limites raisonnables, je me contenterai d'évoquer par des références précises les "gros" théorèmes de la théorie des ensembles analytiques invoqués dans les démonstrations, mais je tâcherai par contre d'expliquer en détail les métamorphoses successives des ensembles semi-polaires (sans toutefois revenir sur les propriétés classiques qu'on trouvera dans [19] et surtout [2]). Un dernier mot avant de commencer : dans toute la suite, "dénombrable" signifie "vide, fini, ou infini dénombrable".

PRELIMINAIRES

Le lecteur pourra, pour se sentir à l'aise, supposer que notre donnée de départ est un semi-groupe de Hunt (et même de Feller) (P_t) sur un espace d'états E localement compact à base dénombrable (en abrégé, LCD), et sa réalisation canonique $(\Omega,\underline{\underline{F}},(\underline{\underline{F}}_t),(\theta_t),(X_t),(P^x))$. En fait, tout ce que nous dirons sur les semi-polaires est encore valable sous les "vieilles" hypothèses droite de Meyer (en abrégé, VHD), auquel cas E est métrisable séparable lusinien, et même, au prix de légers aménagements, sous les "nouvelles" hypothèses droites de Getoor-Sharpe (en abrégé, NHD), auquel cas E est métrisable séparable radonien.

Par convention, <u>toute partie nommée de</u> E <u>sera tacitement supposée presque borélienne</u> (plus précisément, si n'on se trouve sous NHD, presque-borélienne relativement à la topologie de Ray ; cf [16])

Soit F une partie de E. Nous noterons reg F ou F' l'ensemble des points de E réguliers pour F, et dirons que F est <u>finement parfait</u> si reg F = F ; en particulier, $x \varepsilon E$ sera dit finement parfait si $\{x\}$ l'est. Cette notion de fine perfection ne coincide avec la notion analogue attachée à la topologie fine que s'il n'existe aucun point finement parfait en notre sens. En plus du début D_F et du temps d'entrée T_F dans l'ensemble F, nous utiliserons le <u>temps de pénétration</u> C_F dans F défini comme suit : on a $C_F(\omega) \geqq t$ ssi l'ensemble $\{s : 0 < s < t \text{ et } X_s(\omega) \varepsilon F\}$ est dénombrable ; autrement dit, C_F est le début de l'ensemble des points de condensation de l'ensemble aléatoire $\{(t,\omega) : X_t(\omega) \varepsilon F\}$. Ce dernier sera noté I(F) et appelé <u>l'ensemble des rencontres temporelles</u> de F.

THEOREME 1.- C_F <u>est un temps d'arrêt terminal exact</u>, <u>et on a</u>

$$C_F \circ \theta_{C_F} = C_F \text{ sur } \{C_F < \infty\}$$

<u>D</u>/ Les propriétés algébriques sont triviales. Le seul point délicat est la mesurabilité de C_F, assurée par IV-111 et 112 de [8] ; c'est une version adaptée au cas probabiliste d'un théorème classique de Mazurkiewicz et Sierpinski en théorie des ensembles analytiques.

Nous dirons que $x \varepsilon E$ est un <u>point de condensation fine</u> de F si l'on a $P^x\{C_F=0\} = 1$ et un <u>point de raréfaction fine</u> si l'on a $P^x\{C_F=0\} = 0$. Bien entendu, la propriété forte de Markov implique qu'un point est forcément de l'un ou l'autre type. On verra plus loin que l'ensemble des $x \in F$ qui sont de raréfaction fine pour F, qui contient évidemment les points de F irréguliers pour F, est "presque" semi-polaire (semi-polaire, sans restriction, sous l'hypothèse (L)).

Soient μ une loi initiale et F une partie de E. Nous dirons que F est μ-polaire si on a $P^{\mu}\{D_F < \infty\} = 0$, que F est μ-semi-polaire si on peut écrire $F = F_1 \cup F_2$ où F_1 est semi-polaire et F_2 μ-polaire, et que F est μ-temporellement dénombrable si on a $P^{\mu}\{C_F < \infty\} = 0$, i.e. si P^{μ}-presque toute coupe de I(F) est dénombrable. Nous verrons que les ensembles μ-semi-polaires sont exactement les ensembles μ-temporellement dénombrables.

THEOREME 2.- L'ensemble des $x \in F$ qui sont de raréfaction fine pour F est μ-temporellement dénombrable pour toute loi initiale μ.

D/ Comme il s'agit d'un simple démarquage d'un raisonnement classique, je ne rentrerai pas dans les détails. La fonction $f : x \to E^x[e^{-C_F}]$ est 1-excessive d'après le théorème 1, et, pour tout n, le temps de pénétration dans $F_n = \{x = f(x) \leqq 1-\frac{1}{n}\}$ est P^{μ}-p.s. égal à $+\infty$ pour toute loi initiale μ. Ainsi, les F_n sont μ-temporellement dénombrables pour toute μ, et donc aussi leur réunion.

REMARQUES.- 1) Disons qu'une partie F de E est presque semi-polaire si elle est μ-semi-polaire pour toute loi initiale μ soit encore si, pour toute μ, il existe un ensemble semi-polaire F_{μ} tel que I(F) et $I(F_{\mu})$ soient P^{μ}-indistinguables. Je ne sais pas s'il existe des ensembles presque semi-polaires qui ne sont pas semi-polaires : c'est un des problèmes qui ont retardé la publication de mon manuscrit de 1979.

2) Supposons l'hypothèse (L) d'absolue continuité vérifiée, et prenons pour μ une probabilité de référence. Alors, les ensembles μ-polaires sont exactement les ensembles polaires, et donc les ensembles μ-semi-polaires sont exactement les ensembles semi-polaires, qui coincident donc ici avec les ensembles presque semi-polaires.

Soient toujours μ une loi initiale et F une partie de E. Nous allons maintenant introduire des concepts à première vue "exotiques" , mais qui mèneront aux résultats les plus fins sur les semi-polaires, en particulier au fait que la connaissance des ensembles μ-polaires, et des points finement parfaits, entraine celle des ensembles μ-semi-polaires, sans autre connaissance sur le semi-groupe (P_t). Nous noterons J(F) et appellerons <u>ensemble des rencontres spatiales</u> de F l'image de I(F) par le processus (autrement dit, on a $J(F) = \{(x,\omega) : \exists t\ X_t(\omega) = x\}$), et nous dirons que l'ensemble F est μ-<u>spatialement dénombrable</u> si P^{μ}-presque toute coupe de J(F) est dénombrable dans E. Il est clair que tout ensemble μ-temporellement dénombrable est μ-spatialement dénombrable, et nous verrons que l'inverse est vrai, une fois écartée une obstruction évidente (il faut supposer que tout point finement parfait de F est μ-polaire). Enfin, nous dirons que F est μ-<u>mince</u> si, pour toute famille (F_i) de parties disjointes de F, l'ensemble $\{i : F_i$ n'est pas μ-polaire$\}$ est dénombrable. Nous verrons, comme application d'un profond théorème de Mokobodzki, que les ensembles μ-minces sont exactement les ensembles μ-spatialement dénombrables, et donc, d'après ce précède, sont exactement les ensembles qui sont union d'un ensemble μ-semi-polaire et d'une réunion dénombrable de points finement parfaits.

CARACTERISATION DES ENSEMBLES μ-SEMI-POLAIRES

On se donne une loi initiale μ et une partie F de E. Pour abréger le langage, nous dirons qu'une mesure (positive σ-finie sur la tribu borélienne de E) est μ-<u>finement diffuse</u> si elle ne charge pas les ensembles μ-semi-polaires.

Nous allons d'abord montrer, en repiquage de ma thèse plus ou moins publiée dans [3], [4] et [5], que les cinq ou six assertions suivantes concernant F sont équivalentes :

(1) F est μ-semi-polaire

(2) F est μ-temporellement dénombrable

($\underline{2}$) I(F) est P^{μ}-indistinguable d'une réunion dénombrable de graphes de temps d'arrêt

(3) Toute partie finement parfaite de F est μ-polaire

(4) Toute partie compacte de F est μ-semi-polaire

(5) F est de mesure nulle pour toute mesure μ-finement diffuse

A cela, j'aurais voulu pouvoir ajouter (autre problème retardateur) l'assertion

($\underline{5}$) F est de potentiel μ-p.p. nul relativement à toute fonctionnelle additive continue

(2) $\Rightarrow$ ($\underline{5}$) est immédiat tandis que ($\underline{5}$) $\Rightarrow$ (3) a été établie par Azéma dans [1] sous l'hypothèse (L) quand μ est une probabilité de référence. Je renonce à aborder ici les problèmes posés par les fonctionnelles additives.

Puis nous montrerons que tout cela est encore équivalent à

(6) F est μ-spatialement dénombrable, et tout point finement parfait de F est μ-polaire

Et, enfin, nous établirons l'équivalence des quatre assertions suivantes concernant F

(a) F est μ-spatialement dénombrable

(b) Il existe une mesure θ telle que toute partie θ-négligeable de F soit nécessairement μ-polaire

(c) F est μ-mince

(d) Il existe une mesure λ telle que F soit λ-négligeable pour toute mesure μ-finement diffuse orthogonale à λ

L'équivalence des assertions (1) à (5) se fera selon le schéma : (1) $\Rightarrow$ ($\underline{2}$) $\Rightarrow$ (3) $\Rightarrow$ (4) $\Rightarrow$ (5) $\Rightarrow$ (2) $\Rightarrow$ (1) . A la suite de quoi, nous établirons : (2) $\Rightarrow$ (6) $\Rightarrow$ (3) . Et nous terminerons par : (a) $\Leftrightarrow$ (b) $\Leftrightarrow$ (c) $\Leftrightarrow$ (d) . C'est la négation de (d) qui nous permettra plus loin de démontrer le résultat cité dans l'introduction.

(1) $\Rightarrow$ ($\underline{2}$) :

C'est un résultat classique de Hunt pour lequel je renvoie le lecteur à sa référence favorite. On verra (2) $\Rightarrow$ ($\underline{2}$) en même temps que (2) $\Rightarrow$ (1).

(2) $\Rightarrow$ (3) :

C'est une conséquence immédiate du théorème suivant, emprunté à Meyer [19], dont la démonstration n'utilise que des moyens "élémentaires" (à part la mesurabilité des débuts).

THEOREME 3.- _Si_ F _est finement fermé et_ μ-_temporellement dénombrable_, _et si_ (F^{α}) _est la suite transfinie_, _indexée par les ordinaux dénombrables_, _des dérivés fins successifs de_ F (i.e. $F^{\alpha+1} = \operatorname{reg} F^{\alpha}$, $F^{\beta} = \bigcap_{\alpha<\beta} F^{\alpha}$ si β est limite), _alors il existe un ordinal dénombrable_ γ _tel que_ F^{γ} _soit_ μ-_polaire_.

_D_/ Soit A la partie de $\mathbb{R}_+ \times \Omega$ telle que, pour tout ω, la coupe $A(\omega)$ soit l'adhérence de la coupe de $I(F)$ selon ω : comme, pour P^{μ}-presque tout ω, $I(F)(\omega)$ est dénombrable et fermé à droite, $A(\omega)$ est fermé et dénombrable pour P^{μ}-presque tout ω. Par récurrence transfinie, on définit alors la suite transfinie (A^{α}) de parties de $\mathbf{R}_+\times\Omega$ telles que, pour tout ω, $A^{\alpha}(\omega)$ soit le α-ième dérivé (au sens de Cantor) de $A(\omega)$, et on vérifie que chaque A^{α} est P^{μ}-indistinguable d'un ensemble optionnel. Il résulte alors de IV-116 de [8] qu'il existe un ordinal dénombrable γ tel que A^{γ} et $A^{\gamma+1}$ soient P^{μ}-indistinguables et donc tel que A^{γ} soit P^{μ}-évanescent. Par ailleurs, une application banale de la propriété forte de Markov montre que, pour tout ordinal dénombrable α, $I(F^{\alpha})$ est inclus, à un ensemble P^{μ}-évanescent près, dans A^{α}. Par conséquent, F^{γ} est μ-polaire.

Nous obtenons comme corollaire un cas particulier, précieux pour la suite, de (2) $\Rightarrow$ (1)

COROLLAIRE.- <u>Si</u> F <u>est finement fermé et</u> μ-<u>temporellement dénombrable</u>, <u>alors</u> F <u>est</u> μ-<u>semi-polaire</u>.

<u>D</u>/ Résulte de l'égalité $F = [\cup_{\alpha<\gamma}(F^{\alpha}-F^{\alpha+1})]\cup F^{\gamma}$

REMARQUE (quelques subtilités que l'on peut passer sans inconvénients).- Quoique cela ne puisse arriver sous hypothèse (L) [facile à voir], ou encore si (P_t) et F sont boréliens [difficile à voir ! repose sur un théorème de Moschovakis dont on peut trouver des illustrations relatives à notre propos dans Feyel [15]], on peut avoir F (presque-borélien) finement fermé et semi-polaire (sans μ-) tel que F^{α} ne soit jamais polaire (sans μ-) pour α dénombrable. L'exemple le plus simple que je connaisse est le suivant : E est égal à $\mathbb{R}\times\mathbb{R}$, (P_t) est le semi-groupe de la translation uniforme parallèlement au premier facteur, et F est obtenu en choisissant dans $[0,1]\times[0,1]$ un compact K universel pour les compacts du premier facteur (i.e., pour tout compact L du premier facteur, il existe y dans le second tel que $L = K(y)$) et en posant $F = \{(x,y) : K(y)$ est dénombrable $\}$. L'ensemble F est coanalytique, non borélien mais quand même presque-borélien, et finement fermé. Je laisse au lecteur la joie de découvrir que F est semi-polaire [ce n'est pas facile, mais je suis près à répondre à tout appel à l'aide]. Et il est bien connu que F^{α} est non vide pour tout α dénombrable.

(3) => (4) :

C'est une conséquence immédiate du théorème suivant, dont la démonstration n'est pas difficile mais utilise cependant les temps de pénétration.

THEOREME 4.- <u>Si</u> F <u>est finement fermé</u>, <u>alors on peut écrire</u> $F = F_1 \cup F_2$ <u>où</u> F_1 <u>est finement parfait et</u> F_2 <u>est</u> ν-<u>semi-polaire pour toute loi initiale</u> ν.

<u>D</u>/ Prenons pour F_1 l'ensemble des points de condensation fine de F et pour F_2 l'ensemble $F-F_1$, et soit f la fonction 1-excessive $x \to E^x[e^{-C_F}]$. D'une part, F_2 est ν-temporellement pour toute ν d'après le théorème 2, et est la réunion de la suite des ensembles finement fermés $F_2^n = \{x : f(x) \leqq 1-\frac{1}{n}\}$: il est donc ν-semi-polaire pour toute ν d'après le corollaire du théorème 3. D'autre part, $F_1 = \{x : f(x) = 1\}$ est finement fermé et, F_2 étant ce qu'il est, a son temps de pénétration P^{ν}-p.s. égal à celui de F pour toute ν : il est donc finement parfait

REMARQUE.- D'après la remarque précédente, il n'est pas possible de démontrer cette version fine du théorème de Cantor-Bendixson par dérivation transfinie sans aller au delà des ordinaux dénombrables (voir cependant dans [15] ce que fait Feyel à l'aide du théorème de Moschovakis). Ce l'est cependant sous hypothèse (L), cf Meyer [19].

(4) => (5) :

C'est une conséquence immédiate de la régularité intérieure des mesures.

(5) => (2) :

C'est une conséquence immédiate du théorème suivant, dont la démonstration nécessite une connaissance approfondie des ensembles aléatoires

THEOREME 5.- Si le temps de pénétration C_F n'est pas P^{μ}-p.s. infinie, il existe une mesure μ-finement diffuse chargeant F.

D/ D'après IV-111 de [8], il existe un processus croissant continu et intégrable (A_t), non P^{μ}-évanescent, tel que, pour P^{μ}-presque tout ω, la mesure $dA_t(\omega)$ soit portée par la coupe de I(F) selon ω. La mesure m définie par

$$m(f) = E^{\mu}[\int_0^{\infty} f(X_t)\, dA_t]$$

pour f borélienne ≥ 0 a alors les propriétés requises.

REMARQUE.- Un énoncé du type IV-111 de [8] est plus simple à démontrer que je ne le pensais dans [5] : comme on dispose d'une loi de probabilité P (ici, P^{μ}), et donc d'ensembles P-négligeables ou P-évanescents, plus d'un théorème profond sur les ensembles analytiques (théorème de Lusin sur les coupes dénombrables, théorème de Mazurkiewicz-Sierpinski sur les coupes non dénombrables, etc.) admet une version plus faible mais plus facile à démontrer, et généralement suffisante pour l'usage probabiliste (on tombera cependant, dans cet article, sur des exceptions plus loin !).

(2) => (1) :

D'après IV-117 de [8], version pour probabiliste du théorème précité de Lusin, il existe une suite (T_n) de v.a. ≥ 0 telle que I(F) soit P^{μ}-indistinguable de la réunion des graphes des T_n [on peut même, grâce à IV-88 de [8], qui est élémentaire, supposer que les T_n sont des temps d'arrêt, mais nous n'en aurons pas besoin]. Définissons une mesure θ par

$$\theta(f) = \sum 2^{-n} E^{\mu}[f(X_{T_n}) 1_{\{T_n<\infty\}}]$$

où f est borélienne ≥ 0 sur E. Il est clair que cette mesure a, relativement à F, la propriété consignée dans l'assertion (b) : toute partie θ-négligeable de F est μ-polaire [j'ai introduit ce type de mesure pour l'étude des semi-polaires dans ma thèse, sans savoir que cela avait déjà été fait bien avant moi par Mme Hervé en théorie "pure" du potentiel]. Comme la classe des ensembles μ-semi-polaires est héréditaire et stable pour les réunions dénombrables, un lemme classique de théorie de la mesure nous permet d'écrire $\theta = \theta_1 + \theta_2$ où θ_1 est μ-finement diffuse et θ_2 est portée par un ensemble μ-semi-polaire. Comme θ est portée par F, θ_1 est forcément nulle : en effet, tout compact inclus dans F est μ-semi-polaire (corollaire du théorème 3), et θ_1 est intérieurement régulière. Donc, $\theta = \theta_2$ est portée par un ensemble μ-semi-polaire G inclus dans F ; mais alors, comme on a $\theta(F-G) = 0$, F-G est μ-polaire, et c'est fini.

Dans la seconde série d'implications, (2) ⇒ (6) est triviale ; (6) ⇒ (2) l'est aussi si les trajectoires sont injectives (par exemple, si (P_t) est le semi-groupe de la chaleur) mais ne l'est pas du tout en général : nous passerons, comme indiqué plus haut, par (6) ⇒ (3) qui nous amènera à utiliser sous sa forme probabiliste le fait que E est un espace de Baire pour la topologie fine [résultat de Brelot, donné comme une curiosité par Meyer [19], que j'ai utilisé pour la première fois dans une étude [6] de la conjecture (c) ⇒(1) en supposant les points semi-polaires ; Feyel en a aussi, indépendamment de moi, révélé la puissance dans [12], [13] et [14] dont je recommande vivement la lecture].

(6) ⇒ (3) :

Profitant du fait que l'ensemble J(F) des rencontres spatiales est à coupes dénombrables dans $E\times\Omega$ comme l'était plus haut, pour (2) ⇒ (1), l'ensemble I(F) des rencontres temporelles dans $\mathbf{R}_+\times\Omega$, nous allons démontrer un peu mieux (avec une petite escroquerie astérisquée qui sera réparée en remarque).

THEOREME 6.- Si F est μ-spatialement dénombrable, on peut écrire $F = F_1 \cup F_2$ où F_1 est une réunion dénombrable de points finement parfaits alors que toute partie finement parfaite de F_2 est μ-polaire.

D/ D'après III-80 de [8], E est boréliennement isomorphe à un borélien de $\mathbf{R}_+$ [c'est élémentaire si E est LCD, mais je tiens à traiter en même temps le cas où on serait sous VHD ; on verra en remarque ce qu'il convient de faire sous NHD]. Alors, IV-177 de [8] nous permet* d'affirmer que J(F) est contenu, à un ensemble P^{μ}-évanescent près, dans la réunion des graphes d'une suite (S_n) de v.a. à valeurs dans E. La mesure θ sur E définie en posant $\theta(f) = \sum 2^{-n} E^{\mu}[f(S_n)]$ pour toute fonction borélienne f $\geqq 0$ a la propriété vue plus haut : toute partie θ-négligeable de F est μ-polaire. On prend alors pour F_1 un représentant de l'ess. sup. relativement à θ de la famille des points finement parfaits de F, et le reste de l'énoncé résulte du lemme suivant (où la notation F a évidemment changé de sens).

LEMME.- Si F est finement parfait, μ-spatialement dénombrable et non μ-polaire, alors F contient un point finement parfait non μ-polaire.

d/ Il existe un ensemble P^{μ}-négligeable N tel que, pour tout $\omega \in N^c$, la coupe de J(F) selon ω soit dénombrable et que la coupe de I(F) selon ω soit un fermé droit d'adhérence (ordinaire) parfaite (cf T68 de [19]) ; comme F n'est pas μ-polaire, l'ensemble $M = N^c \cap \{D_F < \infty\}$ n'est pas P^{μ}-négligeable. Fixons ω dans M : $I(F)(\omega)$, qui n'est pas loin d'être un parfait droit non vide de $\mathbf{R}_+$, a pour image par $t \to X_t(\omega)$ l'ensemble $J(F)(\omega)$ qui est dénombrable dans E. Comme $I(F)(\omega)$ est un fermé droit et que la topologie droite est une topologie de Baire, on en déduit l'existence d'un $x \in J(F)(\omega)$ et d'un intervalle non vide $]u,v[$ de $\mathbf{R}_+$ tels qu'on ait $X_t(\omega) = x$ pour tout t dans $J(F)(\omega) \cap]u,v[$, et, $J(F)(\omega)$ étant d'adhérence parfaite, l'ensemble $J(F)(\omega) \cap]u,v[$ n'est pas dénombrable [il nous suffit en fait qu'il ait au moins deux points]. Faisons maintenant varier ω. Définissons un temps d'arrêt T par

$$T(\omega) = \inf\{t : X_t(\omega) \in F \text{ et } X_t(\omega) \neq X_0(\omega)\}$$

puis, pour tout $r \in \mathbf{Q}_+$, deux t.d'a. U_r et V_r par

$U_r(\omega) = \inf\{t \geq r : X_t(\omega) \in F\}$, $V_r(\omega) = U_r(\omega) + (T \circ \theta_{U_r})(\omega)$
D'après ce qui précède, pour $\omega \in M$, on est assuré qu'il existe $r(\omega)$ tel que $U_r(\omega) < V_r(\omega)$, et comme on a $P^\mu(M) > 0$, il existe $r \in \mathbf{Q}_+$ tel qu'on ait $P^\mu\{U_r < V_r\} > 0$. Nous fixons un tel r et nous définissons un dernier t.d'a. S par $S = U_r$ sur $\{U_r < V_r\}$ et $S = +\infty$ ailleurs. On a $X_S \in F$ et $T \circ \theta_S > 0$ sur $\{S < +\infty\}$, et la propriété de Markov forte entraine alors l'existence d'un $x \in F$ tel que $P^x\{T>0\}$ soit >0. Comme, F étant finement parfait, x est régulier pour F, la définition de T implique que x est régulier pour lui-même : c'est fini.

REMARQUES.- 1)(au sujet de l'escroquerie) Ω étant muni de la tribu $\underline{\underline{F}}^\mu$ et F étant presque-borélien dans E, I(F) est bien P^μ-indistinguable d'un élément de $\underline{\underline{B}}(\mathbf{R}_+) \times \underline{\underline{F}}^\mu$, mais J(F), qui est image directe de I(F) par $(t,\omega) \to (X_t(\omega),\omega)$ est en général seulement P^μ-indistinguable d'une partie $\underline{\underline{B}}(\mathbf{R}_+) \times \underline{\underline{F}}^\mu$-analytique de $E \times \Omega$ si bien que la référence proposée ci-dessus est en toute rigueur insuffisante. Le lecteur qui s'estimerait grugé par les auteurs du traité incriminé peut cependant se reporter à XI-51 de [9], petits caractères compris, écrit pour un ensemble analytique.

2) Sous NHD, le fait que E soit seulement radonien amène des difficultés de mesurabilité, ou, plutôt, de capacitabilité, insurmontables si on les heurte de front. Un bon moyen d'éviter cet étoc est de plonger E dans un espace métrisable compact $\hat{E}$ et de prolonger (P_t) sur E en un semi-groupe $(\hat{P}_t)$ sur $\hat{E}$ en posant, pour f borélienne ≥ 0 sur $\hat{E}$,

$\hat{P}_t(x,f) = P_t(x,f_{|E})$ pour $x \in E$, $\hat{P}_t(x,f) = f(x)$ pour $x \in \hat{E}-E$
Un exercice proposé par Sharpe dans son livre bien connu assure que $(\hat{P}_t)$ vérifie NHD ; l'espace canonique que l'on a

envie de lui associer est évidemment, en confondant les points de $\hat{E}$-E avec les applications constantes de $\mathbf{R}_+$ dans $\hat{E}$-E, l'espace somme $\hat{\Omega} = \Omega \cup (\hat{E}-E)$ muni de ses applications coordonnées $(\hat{X}_t)$ et des tribus qu'elles engendrent (comme on a $\omega \in \Omega$ ssi $\forall r \in \mathbf{Q}_+$ $\hat{X}_r(\omega) \notin \hat{E}-E$, Ω est universellement mesurable dans $\hat{\Omega}$), et des mesures $\hat{P}^x$ où $\hat{P}^x$ est concentrée sur Ω et égale à P^x pour $x \in E$, est égale à ε_x pour $x \in \hat{E}-E$. Dans ces conditions, $\hat{E}$-E est μ-polaire pour toute loi initiale μ portée par E si bien que, en remplaçant la considération de (P_t) sur E par celle de $(\hat{P}_t)$ sur $\hat{E}$, on ne change guère la situation probabiliste tout en se plaçant dans une bien meilleure situation "analytique" dans laquelle on peut appliquer XI-51 de [9].

Nous passons à la dernière série d'implications. Parmi elles, (a) $\Rightarrow$ (b) a été vue au début de la démonstration du théorème 6, (b) $\Rightarrow$ (c) est triviale, et (c) $\Rightarrow$ (d) résulte de (c) $\Rightarrow$ (b) $\Rightarrow$ (a) [en effet, d'après le théorème 6, F est alors la réunion d'un ensemble μ-semi-polaire et d'une suite (x_n) de points si bien qu'on peut prendre $\lambda = \sum 2^{-n} \varepsilon_{x_n}$]. Il nous reste donc à établir (d) $\Rightarrow$ (c) $\Rightarrow$ (b) $\Rightarrow$ (a), ce que nous ferons après quelques préliminaires destinés à faciliter l'accès aux références.

D'abord, quitte à changer de manière anodine pour nous la situation de départ, nous supposons désormais que E est métrisable compact (on est bien plus à l'aise en théorie des capacités si on évolue dans des espaces métrisables compacts). Si E ne l'était pas, on procéderait comme dans la remarque précédente : plongement de E dans un espace métrisable compact $\hat{E}$ (si E était LCD, on retrouve le célèbre point δ), puis prolongement de (P_t) sur E en $(\hat{P}_t)$ sur $\hat{E}$ en posant $\hat{P}_t f = P_t(f_{|E})$ sur E et $\hat{P}_t f = f$ sur $\hat{E}$-E pour f borélienne ≥ 0 sur $\hat{E}$. Sous Hunt ou VHD, on prend alors pour $\hat{\Omega}$ le nouvel Ω canonique relatif à $\hat{E}$ tandis que, sous NHD, on prend le $\hat{\Omega}$ de la remarque précédente. Dans tous les cas, en oubliant désormais les chapeaux, J(F) est indistinguable d'une partie $\underline{\underline{B}}(E)\times\underline{\underline{F}}^{\mu}$-analytique de l'espace produit $E\times\Omega$, dont le premier facteur est idéal pour la théorie des capacités.

Ensuite, nous identifions l'espace mesurable séparable $(\Omega,\underline{\underline{F}}^{o})$ à une partie de l'espace métrisable compact $W = E^{\mathbb{Q}_+}$ muni de sa tribu borélienne $\underline{\underline{W}}$. Quoique cela soit sans importance pour la suite, rappelons pour fixer les idées ce qu'on sait de la mesurabilité de Ω dans W (cf IV-18 et 19 de [8]) : sous Hunt, Ω, étant l'espace des trajectoires càdlàg, est borélien dans W ; sous VHD, Ω, étant l'espace des trajectoires càd, est coanalytique et donc universellement mesurable dans W ; sous NHD, Ω n'est qu'un sous-ensemble de trajectoires càd, et, en général, on ne peut rien dire sur sa mesurabilité dans W. Dans tous les cas, et c'est ce qui importe pour nous, <u>il existe une partie analytique</u> J <u>de</u> $E\times W$ <u>dont la trace sur</u> $E\times\Omega$ <u>est</u> P^{μ}<u>-indistinguable de</u> J(F). Enfin, nous notons Q la probabilité sur W égale à l'image de P^{μ} par l'injection de Ω dans W ; lorsque Ω est universellement mesurable dans W, il n'y a aucun inconvénient à confondre les mesures P^{μ} et Q.

Nous terminons ces préliminaires avec quelques notions de théorie des capacités. Si X est un espace métrisable compact (par exemple, E ou $E\times W$), nous désignons par X^{+} l'ensemble des mesures de masse ≤ 1 sur X, que nous munissons de la topologie vague : c'est un espace métrisable compact. Cela nous permet de définir une bicapacité γ sur $X^{+}\times X$ en posant pour toute partie analytique H de X^{+} et toute partie analytique A de X

$$\gamma(H,A) = \sup_{\nu\in H} \nu(A)$$

γ est appelée une bicapacité parce qu'elle est croissante et montante en ses deux arguments, et qu'elle est descendante en ses deux arguments quand ceux-ci sont astreints à être compacts. Lorsque H est fixé, compact, on obtient une capacité γ_H sur X [on sait - cf [11] - que toutes les capacités fortement sous-additives sont de ce type] ; lorsque H, fixé, est seulement analytique, γ_H n'est plus en général une capacité (perte de la descente sur les compacts), mais reste néanmoins une excellente fonction d'ensemble : nous dirons que c'est une <u>sous-mesure analytique</u> pour rappeler que c'est un sup (ponctuel) de mesures parcourant une partie analytique de X^{+}. Toutes les fonctions d'ensembles qu'on rencontre en théorie des processus de Markov ou du potentiel sont de ce type (du moins, après aménagement de la situation de départ). Nous allons voir cela à propos de la fonction d'ensemble qui nous intéresse ici.

THEOREME 7.- <u>La fonction d'ensemble</u> C <u>définie par</u>

$$C(A) = P^{\mu}\{D_{A\cap F} < \infty\}$$

<u>pour</u> A <u>analytique dans</u> E <u>est une sous-mesure analytique</u>.

<u>D/</u> Rappelons que nous désignons par J une partie analytique de E×W dont la trace sur E×Ω est P^{μ}-indistinguable de J(F), et par Q la mesure P^{μ} "vue" sur W. Soit L la partie de $(E\times W)^{+}$ constituée des probabilités portées par J et de marge $\leq Q$ sur W : L est intersection d'un analytique et d'un compact,

et est donc analytique dans $(E\times W)^{*}$. Si p désigne la projection de E×W sur W, on a évidemment

(∗) $$Q[p(B\cap J)] \geqq \sup_{\lambda\in L} \lambda(B)$$

pour toute partie analytique B de E×W, et il résulte immédiatement du théorème de section mesurable (cf par exemple III-81 de [8]) que l'inégalité (∗) est en fait une égalité. Soit alors H l'image de L par [l'application de $(E\times W)^{*}$ sur E^{*} induite par] la projection de E×W sur E : H est analytique dans E^{*}, et on vérifie sans peine que, pour toute partie analytique A de E, on a $C(A) = \sup_{\nu\in H} \nu(A)$.

Nous pouvons maintenant revenir à la démonstration des implications restantes, soit (d) ⇒ (c) ⇒ (b) ⇒ (a), et nous commençons par (c) ⇒ (b) qui, avec nos notations et le jargon de [9], signifie que la sous-mesure analytique C est dominable par une mesure sur E si F est μ-mince (ou, plutôt, C-mince d'après [9]). Cela résulte de XI-47 de [9], mais, dans notre situation, c'est plus simple que cela en a l'air : c'est un simple avatar du théorème de section mesurable (nous laissons au lecteur le plaisir de s'en rendre compte lui-même). Passons à (d) ⇒ (c) ou, plutôt, à sa contraposée : c'est une conséquence de XI-46-a) de [9] vu la définition de l'épaisseur adoptée en XI-43 dans [9] (on prendra garde que cette définition de l'épaisseur n'est pas la même que dans [11], même si, en fin de compte, c'est la même chose pour les sous-mesures analytiques ; cela est d'ailleurs signalé au XI-46-a) de [9]). Il ne nous reste plus à établir que l'implication (b) ⇒ (a) qui est, à mon avis, la plus profonde du point de vue de la théorie des processus. C'est une conséquence immédiate du point b) du théorème suivant de Mokobodzki, pour lequel nous donnerons des références après l'énoncé

THEOREME 8.- <u>Soient</u> W <u>et</u> E <u>deux espaces métrisables compacts</u>, Q <u>une probabilité sur</u> W <u>et</u> J <u>une partie analytique de</u> E×W. <u>Posons pour toute partie analytique</u> A <u>de</u> E <u>et tout</u> w∈W

$$C_w(A) = 1 \text{ (resp } 0) \text{ si } A \cap J(w) \neq \emptyset \text{ (resp } = \emptyset)$$

<u>où</u> J(w) <u>est la coupe de</u> J <u>selon</u> w, <u>puis</u>

$$C(A) = \int C_w(A)\, Q(dw)$$

<u>Alors, il existe sur</u> E <u>une probabilité</u> θ <u>vérifiant</u>

a) $\forall \varepsilon > 0 \ \exists \delta > 0 \quad \theta(A) < \delta \Rightarrow C(A) < \varepsilon$

<u>ssi</u> J <u>est</u> Q-<u>p.s. à coupes finies</u>

b) $\theta(A) = 0 \Rightarrow C(A) = 0$

<u>ssi</u> J <u>est</u> Q-<u>p.s. à coupes dénombrables</u>.

Contrairement à ce qu'on pourrait croire, la démonstration de ce théorème n'a pas été le résultat d'une "commande" de ma part à Mokobodzki (il en sera autrement dans la suite de l'exposé !) pour me permettre de démontrer (b) ⇒ (a). La démonstration originale [20], qui n'est pas d'un abord facile, de Mokobodzki résolvait un problème posé par Horowitz et inspiré de résultats classiques de Lusin et Banach lorsque J est un graphe. Il m'a fallu plus d'un an pour me rendre compte que ce théorème me permettait de résoudre la conjecture [6] : le temps nécessaire pour passer des ensembles temporellement dénombrables aux ensembles spatialement dénombrables. Depuis, Feyel, Mokobodzki et moi avons publié une extension de ce théorème dans [11], avec des démonstrations plus transparentes (bien que [11] s'occupe surtout de capacités aléatoires et que je n'ai toujours pas rédigé la suite promise pour les mesures analytiques aléatoires ; dans le cas du théorème ci-dessus, XI-52 de [9] permet de passer d'un analytique J à coupes compactes à un analytique J à coupes quelconques). Par ailleurs, Talagrand a donné une toute autre démonstration dans [21], de nature probabiliste, reprise dans [22] pour atteindre le beau résultat quantitatif de Feyel [11] ; la méthode probabiliste de Talagrand a été simplifiée et amplifiée par Louveau [18] dont certains résultats vont au delà de ceux de [11].

NOYAUX PROPRES IRRATIONNELS

Cette dernière section différera notablement de la précédente : d'abord, elle sera plus courte, et puis, elle ne fera pas appel à la théorie des ensembles analytiques (sauf par le truchement des équivalences précédemment établies), et enfin, elle sera consacrée plutôt à des curiosités intéressantes qu'à des résultats fondamentaux (que l'on espère intéressants !). Un point commun, cependant : ce sont des résultats fins. Par exemple, ici, nous ferons appel non seulement à la plus difficile des équivalences obtenues, mais aussi à la théorie ergodique des résolvantes.

Nous commençons par expliquer le titre. Etant donné un espace mesurable $(E,\underline{\underline{E}})$ et un noyau V (sous-entendu mesurable positif sur $(E,\underline{\underline{E}})$), nous dirons qu'une fonction f sur E est V-intégrable si elle est $\underline{\underline{E}}$-mesurable et si $V(|f|)$ est bornée, et que le noyau V est propre s'il existe une fonction V-intégrable strictement positive. On sait que, si V vérifie le principe complet du maximum (en abrégé, le PCM), cette notion forte de propreté est équivalente à d'autres usuelles, plus faibles. Le noyau V, supposé propre, sera dit <u>rationnel</u> s'il existe deux noyaux bornés A et B, non nécessairement positifs (en abrégé, n.n.p.), avec A non nul, tels qu'on ait

$$AV - B = 0$$

i.e. $AVf - Bf = 0$ pour toute f V-intégrable. La plupart des noyaux propres qu'on rencontre en théorie des processus de Markov ou du potentiel sont rationnels : par exemple, le noyau potentiel (élémentaire) G d'un noyau sous-markovien N vérifie $G = I + NG$ et donc $(I-N)G - I = 0$ s'il est propre ; de même, le noyau potentiel U d'une résolvante sous-markovienne (U_p) vérifie $U = U_1 + U_1U$ et donc $(I-U_1)U - U_1 = 0$ s'il est propre. On aura deviné que le noyau propre V est <u>irrationnel</u> s'il n'est pas rationnel, autrement dit si l'égalité $AV - B = 0$ où A et B sont des noyaux bornés n.n.p. n'est possible que pour $A = B = 0$. Notre but final est de montrer que, sauf cas trivial, pour toute bonne théorie de potentiel, de noyau potentiel U propre (éventuellement borné), il existe une fonction mesurable $\phi \geq 0$, finie, telle que, J_ϕ désignant le noyau de multiplication par ϕ, le noyau propre $V = UJ_\phi$, qui vérifie le PCM, soit irrationnel. Et, pour en arriver là, nous montrerons l'existence de quelques mesures finement diffuses intéressantes pour elles-mêmes ; ces mesures devraient, même dans un cadre plus large que celui qui sera le nôtre, avoir des analogues probabilistes sous forme de fonctionnelles additives continues, mais, je l'ai dit plus haut, je ne tenterai pas d'aborder ici (ni sans doute ailleurs !) les problèmes posés par les fonctionnelles additives, ou, plus généralement, les mesures aléatoires homogènes.

Notre situation de départ reste pour le moment celle des préliminaires, sauf que nous supposons de plus que le semi-groupe (P_t) vérifie l'hypothèse (L) ; nous désignerons désormais par λ une mesure σ-finie de référence. Nous continuerons à utiliser la terminologie introduite auparavant, sauf que nous omettrons le préfixe "μ-" quand la loi initiale sera égale, ou plutôt équivalente, à λ : par exemple, on dira que ν est une mesure finement diffuse si elle ne charge pas les semi-polaires (lesquels sont les mêmes que les λ-semi-polaires : il n'y a pas d'ambiguité).

Voici d'abord, sous uneforme un peu plus générale, le résultat énoncé dans l'introduction

THEOREME 9.- Si E n'admet aucun point finement ouvert (par exemple, si λ est diffuse), alors il existe une mesure ν finement diffuse, orthogonale à λ, et chargeant tous les ouverts fins non vides.

D/ Nous allons commencer par montrer que tout ouvert fin non vide porte une mesure finement diffuse non nulle, orthogonale à λ. Fixons un tel ouvert fin G : comme E est un espace de Baire pour la topologie fine, il résulte du théorème 6 et de (1) $\Leftrightarrow$ (3) que G n'est pas spatialement dénombrable (ou, autrement dit, grâce à (1) $\Leftrightarrow$ (6), G n'est pas la réunion d'un ensemble semi-polaire et d'un ensemble dénombrable). Mais alors, d'après (a) $\Leftrightarrow$ (d), il existe une probabilité finement diffuse ν_G portée par G et orthogonale à λ. Associons maintenant, quand G varie, à chaque mesure ν_G obtenue son support fin F_G : comme on est sous hypothèse (L), il existe une suite (F_{G_n}) extraite de la famille des F_G de sorte que

$\bigcup_n F_{G_n}$ soit finement dense dans E. Il n'y a plus qu'à poser $\nu = \sum 2^{-n} \nu_{G_n}$.

REMARQUE.- On peut se demander s'il existe une fonctionnelle additive continue (A_t), de support fin E, telle que $dA_t(\omega)$ soit étrangère à la mesure de Lebesgue pour presque tout ω.

Sans chercher la plus grande généralité, nous nous plaçons désormais sous de bonnes hypothèses de dualité : E est un espace LCD et (P_t) est un semi-groupe de Hunt admettant, par rapport à la mesure de référence λ, un semi-groupe de Hunt dual $(\hat{P}_t)$. Les résolvantes sont notées (U_p) et $(\hat{U}_p)$, et les noyaux potentiels U et $\hat{U}$ sont supposés propres. Rappelons (cf [2]) que (P_t) et $(\hat{P}_t)$ admettent les mêmes ensembles polaires, semi-polaires, ou de potentiel nul, et que les adhérences fine et cofine ne diffèrent que par un semi-polaire.

Je dois à Mokobodzki la démonstration (avec utilisation du théorème précédent) du théorème suivant, bien plus fin que je ne l'imaginais. Sa démonstration est un peu allégée ici par l'usage de notions probabilistes.

THEOREME 10.- Si E n'admet aucun point finement ouvert, il existe une fonction presque-borélienne $h \geq 0$, finie, telle que la mesure σ-finie $h\lambda$ charge infiniement tout ouvert fin (ou cofin) non vide.

D/ D'abord, quitte à considérer $(e^{-pt}P_t)$ et $(e^{-pt}\hat{P}_t)$ pour un $p>0$ au lieu de (P_t) et $(\hat{P}_t)$, on peut supposer que U et $\hat{U}$ sont bornés. Ceci dit, il suffit évidemment de trouver $h \geq 0$ finie telle que $\langle h,f \rangle_\lambda = +\infty$ pour toute fonction excessive bornée f, et, pour des raisons qui apparaitront plus loin, nous chercherons h vérifiant une propriété encore plus forte à savoir $<h,f>_\lambda = +\infty$ pour tout élément f du cône positif épointé C_* de l'algèbre engendrée par les fonctions excessives bornées [il nous faut disposer de suffisamment de fonctions positives càdlàg sur les trajectoires]. Maintenant, pour f appartenant à C_*, on a

$$<U(fh),f>_\lambda = <fh,\hat{U}f>_\lambda \leqq <f,h>_\lambda \|\hat{U}f\|_\infty$$

si bien que h a la propriété voulue dès que U(fh) est infini sur $\{f>0\}$ pour tout $f\varepsilon C_*$ [en fait, les deux propriétés sont équivalentes, et équivalentes à celle de l'énoncé], et c'est sous cette forme que nous allons démontrer le théorème. Nous nous donnons pour ce faire une mesure ν vérifiant les propriétés énoncées au théorème 9 ; comme U est borné et ν finement diffuse, on peut supposer, quitte à remplacer ν par une mesure équivalente, que ν a son potentiel $u = U\nu$ borné. Soit, pour tout $t>0$, D_t le laplacien approché $\frac{1}{t}(I-P_t)$ et considérons la fonction maximale $h = \sup_{t>0} D_t u$: d'après le lemme maximal (cf XII-67G de [10]), h est finie λ-p.p., et nous allons montrer qu'elle convient (quitte à la rendre finie de manière triviale). Fixons $f\varepsilon C_*$; de $h \geqq D_t u$ on tire la majoration forte $U(fh) \gg U(fD_t u)$, d'où $U(fh) \gg U(f\nu)$ en faisant tendre t vers 0 et en supposant établi

$$(*) \qquad \forall g\varepsilon C_* \quad \lim_{t\downarrow 0} U(gD_t u) = U(g\nu) .$$

S'il existait un $c\varepsilon\mathbf{R}_+$ tel que l'ouvert fin $\{U(fh)<c\}\cap\{f>0\}$ ne soit pas vide, le potentiel d'équilibre e de $\{U(fh)\geqq c\}$ serait <1 sur cet ouvert fin, et les fonctions $\phi = 1-e$ et $f\phi$ appartiendraient à C_* ; mais, U vérifiant le PCM, $U(f\phi h)$ bornée par c sur $\{f\phi h>0\}$ est bornée par c partout, et par ailleurs vérifie $U(f\phi h) \gg U(f\phi\nu)$ d'après ce qui précède en y remplaçant f par $f\phi$: il résulterait alors du théorème de dérivation (cf XII-63 de [10]) que $f\phi\nu$ a une densité par rapport à λ, ce qui est impossible d'après le choix de ν. Ainsi, $U(fh)$ est infini sur $\{f>0\}$ pour $f\in C_*$. Terminons par l'établissement de (*), ce que nous ferons de manière probabiliste. Soit $A=(A_t)$ la fonctionnelle additive naturelle admettant $u=U\nu$ pour potentiel : comme ν est finement diffuse, A est continue. Un calcul élémentaire et classique (propriété simple de Markov et théorème de Fubini) donne

$$U(gD_t u) = E^{\cdot}[\int_0^{\infty} dA_s \left(\frac{1}{t}\int_{(s-t)^+}^{s} g(X_v)\,dv\right)]$$

si bien que la limite de $U(gD_t u)$ quand t tend vers 0 vaut $E^{\cdot}[\int_0^{\infty} (g(X_s))_-\,dA_s]$ et est donc égale, A étant continue, au potentiel de la fonctionnelle additive gA, et donc au potentiel de la mesure $g\nu$ d'après un résultat classique de Meyer.

REMARQUES.- 1) Les théorèmes 9 et 10 sont très simples à établir dans le cas du potentiel de la chaleur ; le théorème 9 l'est encore dans le cas du potentiel newtonien, mais je doute fort qu'il en soit de même pour le théorème 10.

2) On peut se demander si on peut choisir la fonction h de l'énoncé de sorte que $\int_0^t h(X_s)\,ds$ soit p.s. infini pour tout $t>0$. On voit aisément que c'est le cas pour le potentiel de la chaleur.

Et, pour finir,

THEOREME 11.- <u>Si</u> E <u>n'admet aucun point finement ouvert</u>, <u>il existe une fonction presque-borélienne</u> $h\geq 0$, <u>finie</u>, <u>telle que le noyau propre</u> $V=UJ_h$ <u>soit irrationnel</u>.

<u>D</u>/ Toutes les fonctions considérées seront presque-boréliennes. On prend pour h une fonction vérifiant les propriétés

du théorème précédent (qui sera appliqué du côté cofin) ; quitte à remplacer h par h+1, on peut supposer $h>0$. Si V était rationnel, il existerait deux mesures bornées α et β, non nécessairement positives, avec $\alpha \neq 0$, telles qu'on ait $<\alpha V,\phi> = <\beta,\phi>$ pour toute fonction ϕ V-intégrable. Posons $u = \hat{U}\alpha$, et soit f une fonction bornée, g une fonction V-intégrable (si bien que fg est aussi V-intégrable) comprise entre 0 et 1. On a

$$<\beta,fg> = <\alpha V,fg> = <uf,gh>_\lambda$$

Prenons $f = \frac{u}{1+|u|}$ si bien que uf est ≥ 0, et faisons croître g vers 1 : comme β est une mesure bornée et λ une mesure positive, on obtient $<\beta,f> = <uf,h>_\lambda$. Mais, comme uf est positive et cofinement continue, le membre de droite ne peut prendre que les valeurs 0 et $+\infty$ tandis que le membre de gauche est nécessairement fini. Etant donné ce qu'est f, comme h est >0, cela ne se peut que si on a $u = 0$ λ-p.p. et donc partout : ainsi, on a $\alpha = 0$, soit une contradiction.

BIBLIOGRAPHIE

[1] AZEMA J. : Une remarque sur les temps de retour. Trois applications (Sem. Proba. VI, L.N. n.258, 35-50, Springer 1972)

[2] BLUMENTHAL R.M., GETOOR R.K. : Markov Processes and Potential Theory (Acad. Press, N.Y. 1968)

[3] DELLACHERIE C. : Ensembles aléatoires I,II (Sem. Proba. III, L.N. n.88, 97-136, Springer, 1969)

[4] Ensembles analytiques, Capacités, Mesures de Hausdorff (L.N. n°295, Springer, 1972)

[5] Capacités et Processus stochastiques (Springer, Heidelberg 1972)

[6] DELLACHERIE C. : Une conjecture sur les ensembles semi-polaires (Sem. Proba. VII, L.N. n.321, 51-57, Springer, 1973)

[7] Sur la caractérisation des noyaux potentiels (Sém. de Th. du potentiel n.9, L.N. n. , , Springer, 1988)

[8] DELLACHERIE C., MEYER P.-A. : Probabilités et Potentiel Tome I. Ch. I à IV (Hermann, Paris 1975)

[9] Probabilités et Potentiel. Tome III. Ch. IX à XI (Hermann, Paris 1983)

[10] Probabilités et Potentiel. Tome IV. Ch. XII à XVI (Hermann, Paris 1987)

[11] DELLACHERIE C., FEYEL D., MOKOBODZKI G. : Intégrales de capacités fortement sous-additives [avec un appendice de Dellacherie](Sem. Proba. XVI, L.N. n.920, 8-29 (et 29-40), Springer, 1982)

[12] FEYEL D. : Ensembles singuliers associés aux espaces de Banach réticulés (Ann. Inst. Fourier 31, 195-223, 1981)

[13] Remarques sur un résultat de Choquet (Sém. de Th. du potentiel n.6, L.N. n.906 118-125, Springer, 1982)

[14] Sur la théorie fine du potentiel (Bull. Soc. Math. France, 111, 41-58, 1983)

[15] Quelques applications d'un théorème de Moschovakis à la théorie du potentiel (Théorie du Potentiel, Colloque d'Orsay, L.N. n.1096, 280-289, Springer, 1984)

[16] GETOOR R.K. : Markov Processes : Ray processes and right processes (L.N. n. 440, Springer, 1975)

[17] HANSEN W. : Semi-polar sets and quasi-balayage (Math. Ann. 257, 495-517, 1981)

[18] LOUVEAU A. : Minceur et continuité séquentielle des sous-mesures analytiques fortement sous-additives (Sém. Initiation à l'analyse, Publ. de l'Univ. P. et M. Curie n.66, 9p, 23e année, 1983/84)

[19] MEYER P.-A. : Processus de Markov (L.N. n.26, Springer, 1967)

[20] MOKOBODZKI G. : Ensembles à coupes dénombrables et capacités dominées par une mesure (Sem. Proba. XII, L.N. n.649, 491-508, Springer 1978)

[21] TALAGRAND M. : Sur deux résultats de Mokobodzki concernant les ensembles à coupes dénombrables [réferences manquantes]

[22] Sur les résultats de Feyel concernant les épaisseurs (Sem. Proba. XII, L.N. n.649, 1-7, Springer, 1978)

Claude DELLACHERIE
Mathématiques, U.A. 759
Université de Rouen
B.P. n.67
76130 MONT SAINT AIGNAN
FRANCE

VECTOR VALUED STOCHASTIC PROCESSES III

PROJECTIONS AND DUAL PROJECTIONS

by

Nicolae Dinculeanu

1. Introduction.

In the first part of this paper we study the optional and predictable projections of vector valued processes with separable range. The existence of the projections of measurable processes is proved in [1]. Moreover, in [1] we proved that the property of being right (respectively left) continuous is inherited by the optional (respectively predictable) projection, provided that the process has relatively compact range. In this paper we prove the existence of "weak projections" of weakly measurable processes and show that the weak right (resp. left) continuity is inherited by the weak optional (resp. predictable) projection, without restrictions on the range.

In the second part of the paper we prove the existence of dual projections for processes with values in a Banach

AMS Subject classification: Primary 60G07; Secondary 60G20

Keywords and phrases: Banach space, Stochastic process, stochastic measure, finite variation, optional, predictable, projections, dual projections space with the Radon Nikodym Property (RNP).

In an appendix we list some properties of operator valued functions which are measurable with respect to a σ-algebra, in the absence of any measure.

Throughout the paper $(\Omega, \mathcal{F}, P)$ is a probability measure space, $(\mathcal{F}_t)_{t \in R_+}$ is a filtration satisfying the usual conditions, and E, F, G, Z are Banach spaces with norm denoted by $\|\cdot\|$. We assume that $Z \subset F'$ and Z is norming for F, that is, for every $y \in F$ we have $\|y\| = \sup\{|\langle y,z\rangle| ; z \in Z, \|z\| \leq 1\}$.

2. Projections of vector valued processes

In this section we define and prove the existence of optional and predictable projections of processes with values in a Banach space E. For this purpose we extend first the definition of conditional expectations for vector valued random variables (r.v.)

Let $f:\Omega \to E$ be a separably valued r.v. and $\mathcal{G}$ a sub σ-field of $\mathcal{F}$. We know that if $f \in L^1_E$ then the conditional expectation $E(f|\mathcal{G})$ exists, as the unique (a.s.) $\mathcal{G}$-measurable, integrable, E-valued r.v. satisfying

$$\int_A f dP = \int_A E(f|\mathcal{G}) dP, \text{ for } A \in \mathcal{G}.$$

Assume now that $E(\|f\| \,|\, \mathcal{G}) < \infty$ a.s. We can still define the extended conditional expectation $E(f|\mathcal{G})$ as the unique (a.s.) $\mathcal{G}$-measurable, E-valued r.v., such that for every $A \in \mathcal{G}$ with $I_A f \in L^1_E$ we have $I_A E(f|\mathcal{G}) \in L^1_E$ and

$$\int_A f\,dP = \int_A E(f|\mathcal{G})\,dP.$$

To prove the existence of $E(f|\mathcal{G})$ we notice that the hypothesis $E(\|f\| \,|\, \mathcal{G}) < \infty$ a.s. implies the existence of an increasing sequence (A_n) from $\mathcal{G}$, with union Ω, such that $I_{A_n} f \in L^1_E$ for each n (see [4], II. 39). Then, for each n, the conditional expectation $E(I_{A_n} f|\mathcal{G})$ is defined and satisfies

$$I_{A_n} E(I_{A_{n+1}} f|\mathcal{G}) = E(I_{A_n} f|\mathcal{G}) \text{ a.s.}$$

The limit $E(f|\mathcal{G}) = \lim E(I_{A_n} f|\mathcal{G})$ is defined a.s. If $A \in \mathcal{G}$ satisfies $I_A f \in L^1_E$ then by Lebesgue's dominated convergence theorem we have $I_A E(f|\mathcal{G}) \in L^1_E$ and $\int_A f\,dP = \int_A E(f|\mathcal{G})\,dP$.

From the above, the expression "the conditional expectation $E(f|\mathcal{G})$ exists" is equivalent to "$E(\|f\| \,|\, \mathcal{G}) < \infty$ a.s."

Lemma 1. Let $f:\Omega \to L(E,F)$ be a (not necessarily separably valued) function such that $\langle fx,z\rangle$ is $\mathcal{G}$-measurable for every $x \in E$ and $z \in Z$. Assume there is a positive, finite, $\mathcal{G}$-measurable function ϕ such that $\|f\| < \phi$ a.s.

Let $g:\Omega \to E$ be a separably valued r.v. such that $E(g|\mathcal{G})$ exists.

Then for every $z \in Z$, the conditional expectation $E(\langle fg,z\rangle|\mathcal{G})$ exists and we have

$$E(\langle fg,z\rangle|\mathcal{G}) = \langle fE(g|\mathcal{G}),z\rangle \text{ a.s.}$$

<u>Proof</u>. For each $z \in Z$, $\langle fg,z\rangle$ is measurable, by Corollary 19 in the Appendix, and we have

$$|\langle fg,z\rangle| \leq \phi\|g\|\|z\|,$$

hence $E(|\langle fg,z\rangle| \,|\, \mathcal{G}) \leq \phi\|z\|E(\|g\| \,|\, \mathcal{G}) < \infty$ a.s. therefore $E(\langle fg,z\rangle \,|\, \mathcal{G})$ is defined.

To prove the equality, assume first that ϕ is bounded. Then the equality is easily verified for step functions g, and remains valid for $g \in L^1_E$ since both members of the equality are continuous linear mappings of L^1_E into L^1.

If ϕ and g are as in the statement, we set $A_n = \{\phi \leq n\}$ and $B_n = \{E(\|g\| \,|\, \mathcal{G}) \leq n\}$. Then $A_n, B_n \in \mathcal{G}$, $\bigcup A_n = \bigcup B_n = \Omega$ and $I_{A_n}\phi \leq n$, and $I_{B_n}g \in L^1_E$ for each n. Let $C_n = \bigcup\{A_i \cap B_j;\ i,j \leq n\}$. Then (C_n) is a sequence from $\mathcal{G}$ with union Ω and for each n, the function $I_{C_n}\phi$ is bounded and $I_{C_n}g \in L^1_E$. By the first part of the proof, for each n we have

$$I_{C_n}E(\langle fg,z\rangle \,|\, \mathcal{G}) = I_{C_n}\langle fE(g \,|\, \mathcal{G}),z\rangle \text{ a.s.,}$$

hence, the equality in the statement of the lemma follows.

The extended definition of the conditional expectation allows an extended definition of optional and predictable projections.

We notice first that if X is a positive, measurable process, then its optional projection ${}^oX$ satisfies ${}^oX < \infty$ outside an evanescent set iff

$$ {}^{o}X_T I_{\{T<\infty\}} = E(X_T I_{\{T<\infty\}} | \mathscr{F}_T) < \infty \text{ a.s.} $$

for every stopping time T. Similarly, the predictable projection ${}^{p}X$ satisfies ${}^{p}X < \infty$ outside an evanescent set iff

$$ {}^{p}X_T I_{\{T<\infty\}} = E(X_T I_{\{T<\infty\}} | \mathscr{F}_{T-}) < \infty \text{ a.s.} $$

for every predictable stopping time T([4], VI, 44f).

Definition 2. Let $X: R_+ \times \Omega \to E$ *be a measurable, separably valued process, such that* ${}^{o}\|X\| < \infty$ (*resp.* ${}^{p}\|X\| < \infty$) *outside an evanescent set.*

An E-*valued optional (resp. predictable) process* Y *is called the optional (resp. predictable) projection of* X *if for every stopping time* T *we have*

$$ Y_T I_{\{T<\infty\}} = E(X_T I_{\{T<\infty\}} | \mathscr{F}_T) \text{ a.s.} \tag{2.1} $$

(*respectively, for every predictable stopping time* T *we have*

$$ Y_T I_{\{T<\infty\}} = E(X_T I_{\{T<\infty\}} | \mathscr{F}_{T-}) \text{ a.s.}) \tag{2.2} $$

The optional projection and the predictable projection of X *are denoted respectively with* ${}^{o}X$ *and* ${}^{p}X$.

Remarks. 1° The assumption ${}^{o}\|X\| < \infty$ outside an evanescent set is equivalent to

$$ E(\|X\|_T I_{\{T<\infty\}} | \mathscr{F}_T) = {}^{o}\|X\|_T I_{\{T<\infty\}} < \infty $$

a.s. - for every stopping time T; hence the extended conditional expectation $E(X_T I_{\{T<\infty\}} | \mathcal{F}_T)$ is defined. Similarly, the assumption ${}^p\|X\| < \infty$ outside an evanescent set implies the existence of the extended conditional expectation $E(X_T I_{\{T<\infty\}} | \mathcal{F}_{T-})$ for every predictable stopping time T.

2°. The conditions ${}^o\|X\| < \infty$ and ${}^p\|X\| < \infty$ are, evidently satisfied, if X is bounded. They are also satisfied if X is an E-valued cadlag martingale of the form $X_t = E(X_\infty \mathcal{F}_t)$ for $t > 0$ and some $X_\infty \in L^1_E$. In fact, $\|X\|$ is a cadlag submartingale, closed to the right by the r.v. $\|X_\infty\|$; by the stopping theorem ([4], VI. 10) we have $\|X_T\| \in L^1$ for every stopping time T; then $E(\|X_T\| | \mathcal{F}_T) \in L^1$ and $E(\|X_T\| | \mathcal{F}_{T-}) \in L^1$, therefore $E(\|X_T\| I_{\{T<\infty\}} | \mathcal{F}_T) < \infty$ a.s. and $E(\|X_T\| I_{\{T<\infty\}} | \mathcal{F}_{T-}) < \infty$ a.s.

3°. If the optional or the predictable projection of X is defined, it is determined uniquely up to an evanescent set (using the cross section theorems).

4°. If the optional or the predictable projection of X exists, it takes on values (except on an evanescent set) in the separable closed subspace $E_0 \subset E$ spanned by the range of X.

In fact, the set $A = \{(t,\omega); {}^oX_t(\omega) \notin E_0\}$ is optional and for every stopping time T we have $E(X_T I_{\{T<\infty\}} | \mathcal{F}_T) \in E_0$, hence ${}^oX_T I_{\{T<\infty\}} \in E_0$ a.s.; therefore the set $\{\omega; (T(\omega),\omega) \in A\}$ is negligible. It follows that A is evanescent ([4], IV. 86). Similar proof for the predictable projections.

5°. As in the scalar case ([4], VI, 44d) the

equalities (2.1) and (2.2) can be replaced with the following one:

$$E(Y_T I_{\{T<\infty\}}) = E(X_T I_{\{T<\infty\}}) \tag{2.3}$$

for every stopping time T (respectively for every predictable stopping time T).

6°. Again as in the scalar case, we have $Y = {}^oX$ iff

$$Y_T = E(X_T | \mathscr{F}_T) \text{ a.s.} \tag{2.4}$$

for every <u>bounded</u> stopping time T; and $Y = {}^pX$ iff

$$Y_T = E(X_T | \mathscr{F}_{T-}) \text{ a.s.} \tag{2.5}$$

for every <u>bounded</u> predictable stopping time T.

7°. Let $U:E \to F$ be a continuous linear mapping and assume ${}^oX$ (resp. ${}^pX$) is defined. Then ${}^o(UX)$ (resp. ${}^p(UX)$) is defined and we have

$${}^o(UX) = U({}^oX) \text{ (resp. } {}^p(UX) = U({}^pX)).$$

Also, for every $z \in F'$, ${}^o(\langle UX,z\rangle)$ (resp. ${}^p(\langle UX,z\rangle)$) is defined and we have

$${}^o\langle UX,z\rangle = \langle U({}^oX),z\rangle \text{ (resp. } {}^p\langle UX,z\rangle = \langle U({}^pX),z\rangle).$$

In particular, for every $x' \in E'$, we have

$$^{o}\langle X,x'\rangle = \langle {}^{o}X,x'\rangle \text{ (resp. } {}^{p}\langle X,x'\rangle = \langle {}^{p}X,x'\rangle)$$

8°. We have $\|{}^{o}X\| \leq {}^{o}\|X\|$ and $\|{}^{p}X\| \leq {}^{p}\|X\|$.

Example 1. Let f be an E-valued integrable r.v. Let $X_t = f$ be the process constant in time and (H_t) a cadlag version of the martingale $E(f|\mathcal{F}_t)$ (for the existence of a cadlag version see [1]). Doob's stopping theorems are valid for vector valued cadlag martingales [3]: for every finite stopping time T we have

$$H_T = E(f|\mathcal{F}_T) = E(X_T|\mathcal{F}_T)$$

and for every finite predictable stopping time T we have

$$H_{T-} = E(f|\mathcal{F}_{T-}) = E(X_T|\mathcal{F}_{T-}).$$

It follows that, like in the scalar case, H is the optional projection of X and H_- is the predictable projection of X. In particular, H_- is the predictable projections of H.

Example 2. More generally, if H is an E-valued cadlag, local martingale with separable range, then H_- is the predictable projection of H.

In fact, it can be proved that for every predictable stopping time S, the conditional expectation $E(X_S I_{\{S<\infty\}}|\mathcal{F}_{S-})$ exists and is equal to $X_{S-}I_{\{S<\infty\}}$, that is, for every $A \in \mathcal{F}_{S-}$ with $\int_A \|X_S\| I_{\{S<\infty\}} dP < \infty$, we have

$$\int_A X_S I_{\{S<\infty\}} dP = \int_A X_{S-} I_{\{S<\infty\}} dP.$$

The existence of the projections is stated in the following theorem.

Theorem 3. *Let X be an E-valued, measurable process with separable range.*

1.) *If* ${}^{o}\|X\| < \infty$ *outside an evanescent set, then X has an optional projection* ${}^{o}X$.

2.) *If* ${}^{p}\|X\| < \infty$ *outside an evanescent set, then X has a predictable projection* ${}^{p}X$.

The proof of this theorem is given in [1].

Remarks 1°. According to definition 2 and theorem 3, the condition that X has optional (resp. predictable) projection is equivalent to the condition that ${}^{o}\|X\| < \infty$ (resp. ${}^{p}\|X\| < \infty$) outside an evanescent set.

2°. If X is right (resp. left) continuous, we do not know whether its optional (resp. predictable) projection has the same property.

In the particular case when X has relatively compact range, if X is right continuous or cadlag, then so is ${}^{o}X$; and if X is left continuous or caglad then so is ${}^{p}X$ (see [1]).

The following theorem and its corollary express a property of projections similar to that of conditional expectations.

Theorem 4. Let $H:R_+\times\Omega \to L(E,F)$ be a stochastic function such that $\|H\|$ and $\langle Hx,z\rangle$ are optional (resp. predictable) for every $x \in E$ and $z \in F'$.

Let $X:R_+\times\Omega \to E$ be a measurable, separably valued process having optional (resp. predictable) projection.

Then, for every $z \in F'$, $\langle HX,z\rangle$ is measurable, has optional (resp. predictable) projection and we have

$$^{o}\langle HX,z\rangle = \langle H(^{o}X),z\rangle \quad (\text{resp. } ^{p}\langle HX,z\rangle = \langle H(^{p}X),z\rangle)$$

up to an evanescent set.

Proof. We consider the optional case only. By hypothesis, $\|X\|$ is measurable and $^{o}\|X\| < \infty$ outside an evanescent set. Then, for every $z \in F'$ we have $^{o}|\langle HX,z\rangle| \leq {}^{o}(\|H\|\|X\|\|z\|) = \|H\|\|z\|{}^{o}\|X\| < \infty$ outside an evanescent set (see [3], VI, 44e); since by Corollary 19 in the Appendix $\langle HX,z\rangle$ is measurable, we deduce that $\langle HX,z\rangle$ has optional projection. Also, by Corollary 19, $\langle H(^{o}X),z\rangle$ is optional. Then, for every stopping time T, we have, by lemma 1,

$^{o}\langle HX,z\rangle_T I_{\{T<\infty\}} = E(\langle HX,z\rangle_T I_{\{T<\infty\}} | \mathscr{F}_T) = E(\langle H_T X_T,z\rangle I_{\{T<\infty\}} | \mathscr{F}_T) = \langle H_T E(X_T I_{\{T<\infty\}} | \mathscr{F}_T),z\rangle = \langle H_T(^{o}X)_T I_{\{T<\infty\}},z\rangle = \langle H(^{o}X),z\rangle_T I_{\{T<\infty\}}$. Since both $^{o}\langle HX,z\rangle$ and $\langle H(^{o}X),z\rangle$ are optional, they are indistinguishable ([3], IV. 86).

Corollary 5. Let $H:R_+\times\Omega \to E$ be a weakly optional (resp. weakly predictable) stochastic function, such that $\|H\|$ is optional (resp. predictable) and $X:R_+\times\Omega \to E'$ be a measurable, separably valued process having optional (resp. predictable) projection. Then $\langle H,X\rangle$ is measurable, has optional (resp. predictable) projection and we have

$$^{o}\langle H,X\rangle = \langle H,{}^{o}X\rangle \text{ (resp. } ^{p}\langle H,X\rangle = \langle H,{}^{p}X\rangle).$$

Remark. In theorem 4 and Corollary 5 we can replace the condition that $\|H\|$ is optional (resp. predictable) by the condition $\|H\| \leq \phi$ for some optional (resp. predictable) process ϕ.

The following theorems prove the existence of "weak" projections of stochastic functions which are not necessarily strongly measurable.

Theorem 6. Assume E and Z are separable Banach spaces, that $Z \subset F'$ and that Z is norming for F.

Let $X:R_{+}\times\Omega \to L(E,F)$ be a stochastic function such that $\langle Xx,z\rangle$ is measurable for every $x \in E$ and $z \in Z$.

If $^{o}\|X\| < \infty$ (resp $^{p}\|X\| < \infty$) outside an evanescent set, then there exists a stochastic function $Y:R_{+}\times\Omega \to L(E,Z')$, determined uniquely up to an evanescent set, such that, for every $x \in E$ and $z \in X$, the process $\langle Yx,z\rangle$ is optional (resp. predictable) and satisfies

$$\langle Yx,z\rangle = {}^{o}\langle Xx,z\rangle \text{ (resp. } \langle Yx,z\rangle = {}^{p}\langle Xx,z\rangle).$$

If X is bounded, and if for every $x \in X$ and $z \in Z$ the process $\langle Xx,z\rangle$ is right continuous or cadlag (respectively, left continuous or caglad), then $\langle Yx,z\rangle$ is also right continuous or cadlag (resp. left continuous or caglad) for *every* $x \in E$ *and* $z \in Z$.

Proof. We shall consider the optional case only. Since E and Z are separable, the process $\|X\|$ is measurable, hence the optional projection ${}^{o}\|X\|$ is defined. For $x \in E$ and $z \in Z$ we have $|\langle Xx,z\rangle| \leq \|X\|\|x\|\|z\|$, hence

$${}^{o}|\langle Xx,z\rangle| \leq {}^{o}\|X\|\|x\|\|z\| < \infty$$

outside an evanescent set, therefore the optional projection ${}^{o}\langle Xx,z\rangle$ exists ([4], VI, 44f). We choose a version of ${}^{o}\langle Xx,z\rangle$.

Let E_0 be a countable vector space over the rational field, dense in E, and Z_0 be a countable vector space over the rational field, dense in Z and norming for F. For $x_1,x_2,x \in E_0$, for $z_1,z_2,z \in Z_0$ and for α,β rational we have

1') ${}^{o}\langle X(\alpha x_1 + \beta x_2),z) = \alpha({}^{o}\langle Xx_1,z\rangle) + \beta({}^{o}\langle Xx_2,z\rangle)$

outside an evanescent set $R_+\times N(\alpha,\beta,x_1,x_2,z\rangle)$;

2') ${}^{o}\langle Xx,\alpha z_1 + \beta z_2\rangle = \alpha({}^{o}\langle Xx,z_1\rangle) + \beta({}^{o}\langle Xx,z_2\rangle)$

outside an evanescent set $R_+\times N(\alpha,\beta,x,z_1,z_2)$;

3') $|{}^{o}\langle Xx,z\rangle| \leq {}^{o}\|X\|\|x\|\|z\|$

outside an evanescent set $R_+\times N(x,z)$.

The union N of all the sets $N(\alpha,\beta,x_1,x_2,z)$, $N(\alpha,\beta,x,z_1,z_2)$ and $N(x,z)$ is P-negligible and we define ${}^{o}\langle Xx,z\rangle = 0$ on $R_+\times N$ for all $x \in E_0$ and $z \in Z_0$. Then the relations 1'), 2') and 3') are valid everywhere for $x,x_1,x_2 \in E_0$, $z,z_1,z_2 \in Z_0$ and α,β rational.

Let now $(t,\omega) \in R_+\times\Omega$ and $x \in E_0$. The mapping $z \to {}^{o}\langle Xx,z\rangle_t(\omega)$ is a continuous and linear (for rational scalars) functional on Z_0; it can be extended to a continuous linear functional $Y(t,\omega,x) \in Z'$ satisfying $|Y(t,\omega,x)| \leq {}^{O}\|X\|_t(\omega)\|x\|$. The mapping $x \to Y(t,\omega,x)$ of E_0 into Z' is linear (for rational scalars) and continuous, hence it can be extended to a continuous linear mapping $Y_t(\omega) : E \to Z'$ satisfying

$$|\langle Y_t(\omega)x,z\rangle| \leq {}^{O}\|X\|_t(\omega)\|x\|\|z\| \text{ for } x \in E,\ z \in Z.$$

From the definition of $Y_t(\omega)$ we deduce

$$\langle Y_t(\omega)x,z\rangle = {}^{O}\langle Xx,z\rangle_t(\omega)$$

for all $(t,\omega) \in R_+ \times \Omega$, $x \in E_0$ and $z \in Z_0$. Since $\langle Yx,z\rangle$ is optional for $x \in E_0$ and $z \in Z_0$, it remains optional for all $x \in E$ and $z \in Z$. Moreover, for $x \in E$ and $z \in Z$ we have

$$\langle Yx,z\rangle = {}^{O}\langle Xx,z\rangle$$

outside an evanescent set depending on x and z. In fact, let $x \neq 0$ in E and $z \neq 0$ in Z; let $x_n \to x$ in E with $x_n \in E_0$ and $z_n \to z$ in Z with $z_n \in Z_0$. We can assume that $\|x_n\| \leq 2\|x\|$ and $\|z_n\| \leq 2\|z\|$ for all n. Let T be a stopping time and $A \in \mathcal{F}_T$ such that $\int_A \|X\|_T dP < \infty$. Then

$$\lim I_A\langle Y_T x_n, z_n\rangle = I_A\langle Y_T x,z\rangle$$

and

$$\lim I_A\langle X_T x_n, z_n\rangle = I_A\langle X_T x,z\rangle$$

on $\{T < \infty\}$ and these processes are dominated by $4I_A{}^o\|X\|_T\|x\|\|z\|$ and $4I_A\|X\|_T\|x\|\|z\|$ respectively. By Lebesgue's theorem we deduce that

$$\lim I_A\langle X_T x_n, z_n\rangle I_{\{T<\infty\}} = I_A\langle X_T x, z\rangle I_{\{T<\infty\}}$$

in L^1, hence

$$\lim I_A{}^o\langle Xx_n, z_n\rangle_T I_{\{T<\infty\}} = I_A{}^o\langle Xx, z\rangle_T I_{\{T<\infty\}}$$

in L^1. Since for each n we have

$$I_A\langle Y_T x_n, z_n\rangle = I_A{}^o\langle Xx_n, z_n\rangle_T \quad \text{on } \{T < \infty\}$$

it follows that

$$I_A\langle Y_T x, z\rangle = I_A{}^o\langle Xx, z\rangle_T \quad \text{a.s. on } \{T < \infty\}.$$

We deduce that $\langle Yx, z\rangle = {}^o\langle Xx, z\rangle$ except on an evanescent set ([4], IV. 86).

The uniqueness of Y follows from the fact that E and Z are separable and Z is norming for F.

Assume now that X is bounded and that $\langle Xx, z\rangle$ is right continuous for every $x \in E$ and $z \in Z$. Then, ${}^o\langle Xx, z\rangle$ can be chosen right continuous for every $x \in E$ and $z \in Z$ ([4], VI, 47); hence $\langle Yx, z\rangle$ is right continuous for every $x \in E_0$ and $z \in Z_0$. Consider now the stochastic function $L: R_+ \times \Omega \to (E \times Z)'$, defined by $L(x,z) = \langle Yx, z\rangle$ for $(x,z) \in E \times Z$, with norm $\|(x,z)\| = \|x\| + \|z\|$.

Then $\|L\| \leq \|Y\| \leq \sup \|X\| < \infty$. Since $E_0 \times Z_0$ is dense in $E \times Z$, if $t_n \downarrow t_0$ we can apply the Banach-Steinhaus theorem and deduce that $L_{t_n}(x,z) \to L_{t_0}(x,z)$ for all $(x,z) \in E \times Z$ and $\omega \in \Omega$; that is, $\langle Yx,z \rangle$ is right continuous for all $x \in E$ and $z \in Z$. The assertion concerning the left limits is proved similarly.

3. Optional and predictable processes with finite variation

We start with the following characterization of an optional (resp. predictable) vector valued process with finite variation, proved in [7] (theorem 2), which extends the classical results for scalar valued processes with finite variation ([4],VI,57-59).

Theorem 7. Let $A: R_+ \times \Omega \to G$ *be a right continuous, raw, measurable process, with raw, integrable variation and with separable range. Then A is optional (resp. predictable) iff*

$$E(\int_{[0,\infty)} \phi_s dA_s) = E(\int_{[0,\infty)} {}^o\phi_s (\text{resp.} {}^p\phi_s) dA_s$$

for every bounded, measurable, scalar valued process ϕ.

We shall show below that if $G \subset L(E,F)$, we can replace the scalar valued process ϕ by bounded, measurable processes with separable range $X: R_+ \times \Omega \to E$. For this purpose we prove a more general result.

Theorem 8. Let $A: R_+ \times \Omega \to L(E,F)$ *be a raw, right continuous, stochastic function, with raw, integrable variation* $|A|$, *such that, for every* $x \in E$ *and* $z \in Z$, $\langle Ax,z\rangle$ *is a raw, measurable process with finite variation* $|\langle Ax,z\rangle|$.

Then $\langle Ax,z\rangle$ *is optional (resp. predictable) for every* $x \in E$ *and* $z \in Z$ *iff*

$$E(\langle\int_{[0,\infty)} X_s dA_s, z\rangle) = E([\int_{[0,\infty)} {}^{O}X_s (\text{resp. } {}^{P}X_s) dA_s, z\rangle)$$

for every $z \in Z$ *and every bounded, measurable, separably valued process* $X: R_+ \times \Omega \to E$.

Proof. We shall prove only the optional case. If X is a bounded, measurable, separably valued process with values in E, then its optional projection ${}^{O}X$ is also bounded, measurable, E-valued and with separable range, therefore the above integrals are defined (see [7], theorem 5). If for any such X and $z \in Z$ the above equality is satisfied, then taking $X = \phi x$ with $x \in E$ and ϕ a scalar, bounded, measurable process, we have ${}^{O}X = {}^{O}\phi x$, hence

$$E(\int_{[0,\infty)} \phi_s d\langle A_s x, z\rangle) = E(\langle\int_{[0,\infty)} \phi_s x dA_s, z\rangle)$$

$$= E(\langle\int_{[0,\infty)} {}^{O}\phi_s x dA_s, z\rangle) = E(\int_{[0,\infty)} {}^{O}\phi_s d\langle A_s x, z\rangle)$$

therefore, by the classical property, $\langle Ax,z\rangle$ is optional.

Conversely, assume that $\langle Ax,z\rangle$ is optional for every $x \in E$ and $z \in Z$, and let X be a bounded, measurable, E-valued processes with separable range and prove the equality in the statement of the theorem. Let E_0 be the separable subspace of E generated by the range of X and let $z \in Z$.

Define the stochastic function $B: R_+ \times \Omega \to L(E_0, R) = E_0'$ by $Bx = \langle Ax, z\rangle$ for $x \in E_0$. We deduce that for every $x \in E_0$, Bx is optional and has integrable variation. We shall show that B also has finite variation $|B|$ and that $|B|$ is optional and integrable. For $s < t$ we have

$$\|B_t x - B_s x\| \leq \|A_t - A_s\| \|x\| \|z\|, \text{ for } x \in E_0,$$

therefore $\|B_t - B_s\| \leq \|A_t - A_s\| \|z\|$. From this inequality we deduce that B is right continuous and has finite variation $|B|$ satisfying $|B| \leq |A| \|z\|$. Moreover, since Bx is optional for every $x \in E_0$ and E_0 is separable, we deduce that $|B|$ is optional ([9], theorem 4). Therefore, by the preceding inequality, B has integrable variation. We apply now the Radon-Nikodym theorem in [8] to the processes B and $|B|$: there is a stochastic function $H: R_+ \times \Omega \to L(E_0, R)$ such that $\|H\| \leq 1$, Hx is optional for every $x \in E_0$ and

$$B_t x = \int_{[0,t]} H_s x \, d|B|_s \quad \text{a.s.}$$

Moreover,

$$\int_{[0,\infty)} X_s dB_s = \int_{[0,\infty)} H_s X_s d|B|_s \quad \text{a.s.}$$

and also

$$\int_{[0,\infty)} {}^o X_s dB_s = \int_{[0,\infty)} H_s ({}^o X_s) d|B|_s \quad \text{a.s.}$$

But $H({}^o X) = {}^o(HX)$, by theorem 4; therefore, $|B|$ being optional,

$$E(\int_{[0,\infty)} H_s X_s d|B|_s) = E(\int_{[0,\infty)} H_s ({}^oX_s) d|B|_s).$$

hence

$$E(\int_{[0,\infty)} X_s dB_s) = E(\int_{[0,\infty)} {}^oX_s dB_s).$$

Now

$$\int_{[0,\infty)} X_s dB_s = \langle \int_{[0,\infty)} X_s dA_s, z \rangle$$

and

$$\int_{[0,\infty)} {}^oX_s dB_s = \langle \int_{[0,\infty)} {}^oX_s dA_s, z \rangle$$

therefore

$$E(\langle \int_{[0,\infty)} X_s dA_s, z \rangle) = E(\langle \int_{[0,\infty)} {}^oX_s dA_s, z \rangle).$$

If in the above theorem we assume Ax measurable and separably valued for every $x \in E$, then we can delete z in the equality of integerals:

Theorem 9. Let $A: R_+ \times \Omega \to L(E,F)$ be a raw, right continuous function with raw, integrable variation $|A|$. Assume Ax is measurable and separably valued for every $x \in E$. Then the following two assertions are equivalent:

1.) Ax is optional (resp. predictable) for every

$x \in E$;

2.) For every E-valued, bounded, measurable process with separable range X we have

$$E(\int_{[0,\infty)} X_s dA_s) = E(\int_{[0,\infty)} {}^oX_s (\text{resp.} {}^pX_s) dA_s$$

If, moreover, A is measurable and separably valued, the above two assertions are equivalent to the following ones:

3.) A is optional (resp. predictable);

4.) For every bounded, measurable, scalar valued process ϕ we have

$$E(\int_{[0,\infty)} \phi_s dA_s) = E(\int_{[0,\infty)} {}^o\phi_s (\text{resp.} {}^p\phi_s) dA_s).$$

Proof. We remark first that for every $x \in E$, Ax is a process with integrable variation, therefore the integrals in assertion 2) are defined ([7], theorem 5). If A is measurable and separably valued, then the integrals in assertion 4) are defined ([7], theorem 2). We shall prove the equivalence of the assertions only in the optional case.

Assume Ax is optional for every $x \in E$. Then $\langle Ax,z\rangle$ is optional for every $x \in E$ and $z \in F'$; from the preceding theorem we deduce that, for every E-valued, measurable process X with separable range we have

$$\langle E(\int_{[0,\infty)} X_s dA_s), z\rangle = E(\langle\int_{[0,\infty)} X_s dA_s, z\rangle)$$

$$= E(\langle\int_{[0,\infty)} {}^oX_s dA_s, z\rangle) = \langle E(\int_{[0,\infty)} {}^oX_s dA_s), z\rangle$$

for every $z \in F'$, therefore

$$E(\int_{[0,\infty)} X_s dA_s) = E(\int_{[0,\infty)} {}^{o}X_s dA_s).$$

Conversely, if this equality is true for every X as in assertion 2), then for every $z \in F'$ we have

$$E(\langle\int_{[0,\infty)} X_s dA_s, z\rangle) = E(\langle\int_{[0,\infty)} {}^{o}X_s dA_s, z\rangle)$$

therefore, by the preceding theorem, $\langle Ax, z\rangle$ is optional for every $x \in E$ and $z \in F'$. Since Ax is separably valued, it follows that Ax is optional (see [6] proposition 22, p. 105, or Appendix, Corollary 22). This proves the equivalence of assertions 1) and 2). If A is separably valued, then the equivalence of assertions 1) and 3) follows from ([6], Corollary of proposition 18, p. 103, or Appendix, propositions 20). Finally, the equivalence of assertions 3) and 4) is stated in theorem 7.

Remark. If A is a scalar valued process with integrable variation and E is a Banach space, we can consider that A has separable range in the space L(E,E) and deduce that assertions 2), 3) and 4) are equivalent.

4. Optional and predictable stochastic measures

We shall denote $\mathcal{M} = \mathcal{B}(R_+) \times \mathcal{F}$. A stochastic measure $\mu: \mathcal{M} \to G$ (i.e. vanishing on evanescent sets) with finite variation is called optional (resp. predictable) if $\mu(\phi) = \mu({}^{o}\phi)$ (resp. $\mu(\phi) = \mu({}^{p}\phi)$) for any bounded, measurable, scalar valued process ϕ. This property remains valid for vector valued processes:

Theorem 10. *A stochastic measure* $\mu : \mathcal{M} \to L(E,F)$ *with finite variation is optional (resp. predictable) iff, for every* E-*valued, bounded, measurable process* X *with separable range we have*

$$\mu(X) = \mu({}^{o}X) \quad (\text{resp. } \mu(X) = \mu({}^{p}X)).$$

Proof. If μ is optional and X is as in the statement, then, for every $x \in E$ and $z \in Z$ we have ${}^{o}\langle Xx,z\rangle = \langle {}^{o}Xx,z\rangle$, therefore

$$\langle \mu(X)x,z\rangle = \mu(\langle Xx,z\rangle) = \mu({}^{o}\langle Xx,z\rangle)$$
$$= \mu(\langle {}^{o}Xx,z\rangle) = \langle \mu({}^{o}X)x,z\rangle$$

hence $\mu(X) = \mu({}^{o}X)$. The converse implication is evident. Similar proof for the predictable case.

Definition 11. *For any stochastic measure* $\mu : \mathcal{M} \to G$ *with finite variation we call optional (resp. predictable) projection of* μ *the stochastic measure* μ^{o} (*resp.* μ^{p}) *defined for every bounded, measurable, scalar valued process* ϕ *by*

$$\mu^{o}(\phi) = \mu({}^{o}\phi) \quad (\text{resp. } \mu^{p}(\phi) = \mu({}^{p}\phi)).$$

Remarks.

1.) According to this definition, μ is optional (resp. predictable) iff $\mu = \mu^{o}$ (resp. $\mu = \mu^{p}$).

2.) If $\mu : \mathcal{M} \to L(E,F)$ is a stochastic measure with finite variation and X is an E-valued, bounded, measurable, separably valued process, then

$$\mu^o(X) = \mu({}^oX) \text{ (resp. } \mu^p(X) = \mu({}^pX)).$$

The proof is similar to that of theorem 10.

3.) If we denote by $|\mu|$ the variation of a stochastic measure μ, then we have

$$|\mu^o| \leq |\mu|^o \text{ and } |\mu^p| \leq |\mu|^p.$$

If $|\mu^o|$ is optional, then $|\mu^o| = |\mu|^o$. If $|\mu^p|$ is predictable, then $|\mu^p| = |\mu|^p$. (see [9]).

5. Dual projections of vector valued processes with finite variation

In this section we define the dual projections of vector valued processes with integrable variation and prove the existence of dual projections for processes with values in a space with the Radon-Nikodym property.

Definition 12. Let G be a Banach space and $A: R_+ \times \Omega \to G$ be a raw, right continuous, separably valued process with raw, integrable variation $|A|$.

An optional (resp. predictable) process $B: R_+ \times \Omega \to G$ with separable range and integrable variation is called the optional (resp. predictable) dual projection of A and is denoted by A^o (resp. A^p) if it satisfies the equality

$$E(\int_{[0,\infty)} \phi_s dB_s) = E(\int_{[0,\infty)} {}^o\phi_s (\underline{\text{resp.}}\ {}^p\phi_s) dA_s)$$

for any bounded, measurable, scalar valued process ϕ.

Remark. We have the following inequalities:

$$|A^o| \leq |A|^o \text{ and } |A^p| \leq |A|^p.$$

In fact, if ϕ is a scalar valued, bounded, measurable process, we have

$$|\mu_{A^o}(\phi)| = |\mu_A({}^o\phi)| \leq |\mu_A|(|{}^o\phi|)$$

$$\leq \mu_{|A|}({}^o|\phi|) = \mu_{|A|^o}(|\phi|)$$

hence $|\mu_{A^o}| \leq \mu_{|A|^o}$. But $|\mu_{A^o}| = \mu_{|A^o|}$, hence

$\mu_{|A^o|} \leq \mu_{|A|^o}$ therefore $|A^o| \leq |A|^o$.

We prove first that the equality in definition 2 remains valid for vector valued processes X instead of scalar processes ϕ.

Theorem 13. Let $A: R_+ \times \Omega \to G$ be a raw, right continuous process with separable range and with raw, integrable variation $|A|$. Assume A has optional (resp. proedictable) dual projection B.

Let E, F be Banach spaces such that $G \subset L(E,F)$. Then for any E-valued, bounded, measurable process X with separable range we have

$$E(\int_{[0,\infty)} X_s dB_s) = E(\int_{[0,\infty)} {}^{o}X_s \text{ (resp. } {}^{p}X_s) dA_s)$$

Proof. Assume $B = A^o$. By hypothesis, for every bounded, measurable, scalar valued process ϕ we have

$$\mu_B(\phi) = E(\int_{[0,\infty)} \phi_s dB_s) = E(\int_{[0,\infty)} {}^{o}\phi_s dA_s) = \mu_A({}^{o}\phi) = (\mu_A)^{o}(\phi)$$

hence $\mu_B = (\mu_A)^o$. Then, by Remark 2 following definition 11 we have $\mu_B(X) = (\mu_A)^o(X) = \mu_A({}^{o}X)$ for X as in the statement, hence

$$E(\int_{[0,\infty)} X_s dB_s) = E(\int_{[0,\infty)} {}^{o}X_s dA_s)$$

The case $B = A^p$ is treated similarly.

The existence of dual projections is stated and proved in the following two theorems.

Theorem 14. If G is a Banach space having the Radon-Nikodym property, then any right continuous, raw, separably valued process $A: R_+ \times \Omega \to G$ with raw, integrable variation has optional (resp. predictable) dual projection.

This will follow by applying assertion 2b of the following theorem, with $G = L(R,G)$.

Theorem 15. Let $A: R_+ \times \Omega \to L(E,F)$ be a right continuous stochastic function with finite variation $|A|$, such that

$\langle Ax,z\rangle$ is measurable for every $x \in E$ and $z \in Z$. Assume $|A|$ is dominated by a raw, increasing, integrable process C, that is, $|A|_t - |A|_s \leq C_t - C_s$ for $s < t$. Then

1.) There exists a right continuous, stochastic function $B: R_+ \times \Omega \to L(E,Z')$, with integrable variation, such that $\langle Bx,z\rangle$ is optional (resp. predictable) for every $x \in E$ and $z \in Z$ and satisfying

$$E(\int_{[0,\infty)} \phi_s d\langle B_s x,z\rangle) = E(\int_{[0,\infty)} {}^o\phi_s (\text{resp. } {}^p\phi_s) d\langle A_s x,z\rangle)$$

for every bounded, measurable, scalar valued process ϕ.

1') For every $z \in Z$ and every E-valued, bounded, measurable process X with separable range, we have

$$E(\langle\int_{[0,\infty)} X_s dB_s, z\rangle) = E(\langle\int_{[0,\infty)} {}^oX_s (\text{resp. } {}^pX_s) dA_s, z\rangle).$$

2.) B can be chosen with values in $L(E,F)$ in each of the following cases:

a.) F is a dual of a Banach space G and we take $Z = G$, hence $F = Z'$;

b.) E is separable, F has the Radon-Nikodym property and Ax is separably valued for every $x \in E$; then Bx is optional (resp. predictable) for every $x \in E$. In this case we have

$$E(\int_{[0,\infty)} X_s dB_s) = E(\int_{[0,\infty)} {}^oX_s (\text{resp. } {}^pX_s) dA_s),$$

for every E-valued, bounded, measurable process X with separable range.

Proof. We consider only the optional case. We apply first Remark 2 following theorem 5 in [7] to obtain a stochastic measure $\mu: \mathcal{M} \to L(E,Z')$ with finite variation, satisfying

$$\langle \mu(X), z \rangle = E(\langle \int_{[0,\infty)} X_s dA_s, z \rangle)$$

for $z \in Z$ and $X \in L^1_E(\mu)$. Then we apply theorem 6 in [7] to the optional projection μ^o of μ and obtain a stochastic function $B: R_+ \times \Omega \to L(E,Z')$ such that $\langle Bx, z \rangle$ is optional for every $x \in E$ and $z \in Z$, and such that

$$\langle \mu^o(X), z \rangle = E(\langle \int_{[0,\infty)} X_s dB_s, z \rangle)$$

for $x \in Z$ and $X \in L^1_E(\mu^o)$. Then statement 2') follows from the equality $\mu^o(X) = \mu({}^oX)$ for X as in the statement, using remark 2 following Definition 11.

Assertion 2a is evident. To prove 2b, assume E is separable, F has the Radon-Nikodym property, and Ax is separably valued (hence measurable) for every $x \in E$. Then we can also assume that F is separable. From theorem 5 in [7] we deduce first that μ and μ^o have values in L(E,F). Then we apply theorem 6 in [7] to deduce 2b.

Theorem 16. Let A, B be two measurable, separably valued, right continuous processes with raw, integrable variation, with values in a Banach space G.

a.) Assume A and B have optional dual projections. Then A and B have the same optional dual projection iff

$$E(A_\infty - A_{T-} | \mathcal{F}_T) = E(B_\infty - B_{T-} | \mathcal{F}_T) \text{ a.s.}$$

for any stopping time T (we denoted $A_\infty = A_{\infty-}$, $B_\infty = B_{\infty-}$).

b.) Assume A and B have predictable dual projections. Then A and B have the same predictable dual projection iff

$$E(A_\infty - A_t | \mathcal{F}_t) = E(B_\infty - B_t | \mathcal{F}_t) \text{ a.s.}$$

for all $t > 0$ and $E(A_0 | \mathcal{F}_{0-}) = E(B_0 | \mathcal{F}_{0-})$ a.s.

The proof is the same as in the scalar case ([4], VI, 69, 75), using the measures μ_A and μ_B associated to A and B ([7], theorem 2).

Theorem 17. Let $A: R_+ \times \Omega \to L(E,F)$ be an optional process with integrable variations and separable range. Then A is predictable iff, for any E-valued, cadlag, bounded martingale $H_t = E(H_\infty | \mathcal{F}_t)$ we have

$$E(H_\infty A_\infty) = E\left(\int_{[0,\infty)} H_{s-} dA_s\right).$$

The proof is the same as in the scalar case ([4], VI, 61), using the fact that H_- is the predictable projection of H.

6. Appendix

We list here some measurability properties on a measurable space without a measure. This is the case of the set $R_+ \times \Omega$ endowed with $\mathcal{B}(R_+) \times \mathcal{F}$ or with the σ-algebra of optional (or predictable, or progressive) sets. These properties were used in the preceding theorems and will be used again in other papers; they were proved in [6] for functions which were measurable with respect to a measure.

Let (X,Σ) be a measurable space. The spaces E,F,G are endowed with their Borel σ-fields. For example, we say that $f:X \to E$ is measurable if $f^{-1}(B) \in \Sigma$ for any Borel set $B \subset E$.

We assume that $G \subset F'$ and that G is norming for F, that is, $\|y\| = \sup\{|\langle y,z\rangle|; z \in G, \|z\| < 1\}$ for every $y \in F$.

Proposition 18. *Let* $U:X \to L(E,F)$ *be such that* Ux *is measurable for every* $x \in E$ *and let* $f:X \to E$ *be measurable and separably valued. Then the function* $Uf:X \to F$ *is measurable.*

Proof: The assertion is evidently true for a step function f, then for a countably valued function f (which is a limit of step functions) and finally for a separably valued function f (which is the limit of a sequence of countably valued functions).

Corollary 19. *Let* $U:X \to L(E,F)$ *be such that* $\langle Ux,z\rangle$ *is measurable for every* $x \in E$ *and* $z \in G$. *Let* $f:X \to E$ *and* $h:X \to G$ *be measurable and separably valued. Then the function* $\langle Uf,h\rangle$ *is measurable.*

Corollary 20. *If* $U:X \to L(E,F)$ *is separably valued and* Ux *is measurable for every* $x \in E$, *then* U *is measurable.*

The proof is the same as that of ([6], prop. 18, p.p.

102-103).

Corollary 21. *If* $U:X \to L(E,F)$ *is separably valued and* $\langle Ux,z\rangle$ *is measurable for every* $x \in E$ *and* $z \in G$, *then* U *is measurable*.

Corollary 22. *If* $f:X \to F$ *is separably valued and* $\langle f,z\rangle$ *is measurable for every* $z \in G$, *then* f *is measurable*.

Bibliography

[1] J. K. Brooks and N. Dinculeanu, Projections and Regularity of Abstract Processes, Stochastic Analysis and Applications, 5 (1987) 17-25.

[2] ______, Espaces H^1 et BMO de martingales abstraites, Comptes Rendus de l'Acad. de Paris, 302 (1986), 639-640.

[3] ______, H^1 and BMO spaces of abstract martingales, Seminar on Stochastic Processes, 1985, Birkhauser, Boston, 1986, 9-34.

[4] C. Dellacherie and P. A. Meyer, Probabilities and Potentials, North Holand, 1975, 1980.

[5] J. K. L. Chung and R. J. Williams, Introduction to Stochastic Integration, Birkhäuser, 1983.

[6] N. Dinculeanu, Vector measures, Pergammon Press, Oxford, 1967.

[7] ______, Vector valued stochastic processes I. Vector valued measures and processes with finite variation, J. Theoretical Probability, 1 (1988).

[8] ______, Vector valued stochastic processes II. A Radon-Nikodym theorem for vector valued processes with finite variation, Proc. Amer. Math. Soc.

[9] ______, Vector valued stochastic processes V. Optional and predictable variation of stochastic measures and stochastic processes.

[10] A. and C. Ionescu Tulcea, Topics in the theory of lifting, Springer, 1969.

[11] G. Kallianpur, Stochastic Filtering theory, Springer, 1980.

[12] A. V. Kussmaul, Stochastic Integration and generalized martingales, Pitman, London, 1977.

[13] M. Métivier, Semimartingales, Walter de Gruyter, Berlin, 1982.

[14] M. Métivier and J. Pellaumail, Stochastic Integration, Academic Press, New York, 1980.

[15] J. Neveu, Martingales à temps discret, Masson, Paris, 1972.

N. Dinculeanu
Department of Mathematics
University of Florida
Gainesville, Florida 32611

ON A CONNECTION BETWEEN KUZNETSOV PROCESSES AND QUASI-PROCESSES

by

P. J. Fitzsimmons*

1. Introduction

Let $X = (X_t, P^x)$ be a Borel right process on a Lusin state space $(E, \mathcal{E})$, with semigroup (P_t). Let m be a measure on $(E, \mathcal{E})$ that is excessive for (P_t). Associated with the pair $\{m, P_t\}$ are (at least) two Markov proccsses of interest. Both processes can be realized as the coordinate process on the space W comprising those paths $w: \mathbf{R} \to E \cup \{\Delta\}$ that are E-valued and right continuous on some open interval $]\alpha(w), \beta(w)[$, taking the value Δ outside of $]\alpha(w), \beta(w)[$. (Here $\Delta \notin E$ is a cemetary, and the dead path $[\Delta]: t \to \Delta$ has $\alpha([\Delta]) = +\infty$, $\beta([\Delta]) = -\infty$.) Put $Y_t(w) = w(t)$ and define σ-fields $\mathcal{G} = \sigma\{Y_t : t \in \mathbf{R}\}$, $\mathcal{G}_t = \sigma\{Y_s : s \leq t\}$. The *Kuznetsov process* associated with $\{m, P_t\}$ is the unique measure Q_m on $(W, \mathcal{G})$ that is carried by $W \backslash \{[\Delta]\}$ and satisfies

$$\begin{aligned} &Q_m(\alpha < t_1, Y_{t_1} \in dx_1, \ldots, Y_{t_n} \in dx_n, t_n < \beta) \\ &\qquad = m(dx_1) P_{t_2 - t_1}(x_1, dx_2) \cdots P_{t_n - t_{n-1}}(x_{n-1}, dx_n) \end{aligned} \tag{1.1}$$

for $-\infty < t_1 < \cdots < t_n < +\infty$. Define a family of shift operators on W by

$$(\sigma_t w)(s) = w(t+s), \qquad s, t \in \mathbf{R},$$

* Research supported in part by NSF grant DMS 8419377.

and note that by (1.1), $\sigma_t(Q_m) = Q_m$, $\forall t \in \mathbf{R}$. Thus (Y_t, Q_m) is a *stationary* Markov process with semigroup (P_t), random "life interval" $]\alpha, \beta[$, and one dimensional distributions (when alive) all m. See [7] for the construction of such processes.

In his study [6] of the Martin boundary for Markov chains, Hunt introduced the notion "approximate Markov chain." Weil [10,11] extended this notion to continuous space and time under the name "quasi-process." To describe this process let $\mathcal{A}$ denote the σ-field of (σ_t)-invariant events in $\mathcal{G}$, and call a $(\mathcal{G}_{t+})$-stopping time S *intrinsic* provided $\alpha \le S < \beta$ on $\{S < +\infty\}$, and $S = t + S \circ \sigma_t$, $\forall t \in \mathbf{R}$. The *quasi-process* associated with $\{m, P_t\}$ is the measure $\mathbf{P}_m$ defined on $(W, \mathcal{A})$, carried by $W \backslash \{[\Delta]\}$, and determined by the conditions

$$(1.2) \qquad \mathbf{P}_m\left(\int_{\mathbf{R}} f \circ Y_t \, dt\right) = m(f), \qquad f \in \mathcal{E}^+ \ (f(\Delta) = 0);$$

and

$$(1.3) \qquad \text{for any intrinsic stopping time } S, \ \{Y_{S+t} : t > 0\} \text{ under } \mathbf{P}_m\,|_{\{S \in \mathbf{R}\}} \text{ is Markovian with semigroup } (P_t).$$

(Note that Y_{S+t} is $\mathcal{A}$-measurable if S is intrinsic.)

Our purpose in this note is to point out a simple and useful connection between Q_m and $\mathbf{P}_m$. In making this connection we use a "switching identity" for flows; this result is recorded in the next section, and is related to identities found in [1] and [8]. In section 3 we discuss the quasi-process $\mathbf{P}_m$ and its connection to Q_m. In section 4 we indicate how the identity of section 2 can be used to derive a variety of formulae, including an illuminating formula for capacities.

2. Flows

It follows from (1.1) and σ-finiteness of the excessive measure m that $(W, \mathcal{G}, Q_m)$ is a σ-finite measure space. Also, $(\sigma_t : t \in \mathbf{R})$ is a measurable group of automorphisms of $(W, \mathcal{G}, Q_m)$. In short, the system $(W, \mathcal{G}, \sigma_t, Q_m)$ is a *flow*. In this section we are concerned only with this flow structure. Since m will remain fixed we drop it from the notation, writing Q for Q_m.

Given $F \in \mathcal{G}^+$ write $\overline{F} = \int_{\mathbf{R}} F \circ \sigma_t \, dt$. Then $\overline{F} \in \mathcal{A}^+$, where $\mathcal{A}$ is the σ-field of (σ_t)-invariant events, as before. Everything that follows is based on the identity

$$(2.1) \qquad Q(F \cdot \overline{G}; A) = Q(\overline{F} \cdot G; A)$$

valid for $F, G \in \mathcal{G}^+$ and $A \in \mathcal{A}$. Formula (2.1) follows easily from the (σ_t)-invariance of Q and Fubini's theorem.

(2.2) Definition. *A $\mathcal{G}$-measurable random time $S: W \to \overline{\mathbf{R}} \equiv [-\infty, +\infty]$ is stationary provided $S = t + S \circ \sigma_t$ for all $t \in \mathbf{R}$.*

If S is stationary then $\{S \in \mathbf{R}\} \in \mathcal{A}$, and for any $B \in \mathcal{R}$ (the Borel subsets of $\mathbf{R}$),

$$(2.3) \qquad \overline{1_{\{S \in B\}}} = \text{Leb}(B) \cdot 1_{\{S \in \mathbf{R}\}}.$$

Substituting (2.3) into (2.1) we obtain a proof of (2.5) below:

(2.4) Proposition. *Let S and T be stationary times. For $A \in \mathcal{A}$, $B \in \mathcal{R}$,*

$$(2.5) \qquad Q(S \in B, T \in \mathbf{R}; A) = Q(S \in \mathbf{R}, T \in B; A)).$$

If, in addition, $\{S \in \mathbf{R}\} \subset \{T \in \mathbf{R}\}$ Q-a.s., then

$$(2.6) \qquad Q(S \in B; A) \leq Q(T \in B; A).$$

Proof. For (2.6) use (2.5): $Q(S \in B; A) = Q(S \in B, T \in \mathbf{R}; A) = Q(T \in B, S \in \mathbf{R}; A) \leq Q(T \in B; A)$. ∎

Recall that the (σ_t)-invariant measure Q is *dissipative* if $\overline{F} < +\infty$ Q-a.s. for each $F \in L^1(Q)^+$, and that Q is *conservative* if $\overline{F} = +\infty$ Q-a.s. for each strictly positive $F \in L^1(Q)$.

(2.7) Proposition. *(a) Q is dissipative if and only if there exists a stationary time S with $Q(S \notin \mathbf{R}) = 0$.*

(b) Q is conservative if and only if $Q(S \in \mathbf{R}) = 0$ for each stationary time S.

Proof. We prove only (a); the argument for (b) is similar. First assume that Q is dissipative, and choose $F > 0$ in $L^1(Q)$. Then $Q(\overline{F} = +\infty) = 0$ and so

$$0 < \int_{-\infty}^{t} F \circ \sigma_s \, ds \downarrow 0 \quad \text{as} \quad t \downarrow -\infty, Q\text{-a.s.} \tag{2.8}$$

For each $n \in \mathbf{N}$ define a stationary time $S_n \equiv \inf \{t : \int_{-\infty}^{t} F \circ \sigma_s \, ds > 1/n\}$, and put $S_0 = +\infty$. Now define a stationary time S by

$$\begin{aligned} S &= S_n \quad \text{on} \quad \{\overline{F} < +\infty\} \cap \{S_n < +\infty,\ S_{n-1} = +\infty\},\ n \in \mathbf{N}, \\ &= +\infty \quad \text{on} \quad \{\overline{F} = +\infty\} \cup \{S_n = +\infty,\ \forall n\}. \end{aligned} \tag{2.9}$$

By (2.8), $S_n > -\infty$ and $S_n \downarrow -\infty$ as $n \to \infty$, Q-a.s. Thus $Q(S \notin \mathbf{R}) = 0$ as required. Conversely, suppose that S is a stationary time with $Q(S \notin \mathbf{R}) = 0$. Fix $F \in L^1(Q)^+$ and note that by (2.1) and (2.3),

$$Q(\overline{F}; a < S < b) = (b - a)Q(F; S \in \mathbf{R}) < +\infty.$$

Thus $\overline{F} < +\infty$ Q-a.s. on $\{a < S < b\}$ for any two reals $a < b$. It follows that $\overline{F} < +\infty$ Q-a.s. on $\{S \in \mathbf{R}\}$, hence $Q(\overline{F} = +\infty) = 0$ as desired. ∎

An interesting consequence of (2.7(b)) is a quick proof of the well known fact that a finite invariant measure Q is conservative. For if S is any stationary time then

$$+\infty > Q(W) \geq Q(a < S < b) = Q(0 < S < 1)(b - a). \tag{2.10}$$

Sending $b - a$ to $+\infty$ in (2.10) we see that $Q(0 < S < 1) = 0$. This forces $Q(S \in \mathbf{R}) = 0$, and since S was arbitrary, Q is conservative.

Now assume that Q is dissipative and fix a stationary time S^* with $Q(S^* \notin \mathbf{R}) = 0$. Define a measure $\mathbf{P}$ on $(W, \mathcal{A})$ by

$$\mathbf{P}(A) = Q(A; 0 < S^* \leq 1), \qquad A \in \mathcal{A}. \tag{2.11}$$

By (2.5), $\mathbf{P}$ is unchanged if S^* is replaced by any other stationary time S^{**} satisfying $Q(S^{**} \notin \mathbf{R}) = 0$. As we shall see in the next section, $\mathbf{P}$ is the quasi-process associated with $\{m, P_t\}$.

Using (2.1) we can invert (2.11) to obtain

$$Q(F) = \mathbf{P}(\overline{F}), \quad F \in \mathcal{G}^+. \tag{2.12}$$

Thus $\mathbf{P}$ determines Q uniquely. Conversely, given $\mathbf{P}$, a countable sum of finite measures on $(W, \mathcal{A})$, let a measure Q be defined by (2.12). Then Q is (σ_t)-invariant. If Q is σ-finite than $\mathbf{P}$ is determined by (2.11). We have thus recovered Theorem 1 of Dynkin [2].

3. Quasi-process

Now recall that $(Y_t : t \in \mathbf{R})$ under Q_m is a Markov process with semigroup (P_t). For the rest of the paper we assume that m is dissipative: $\int_0^\infty P_t f \, dt < \infty$ m-a.e. for each $f \in L^1(m)^+$. This is the case if and only if Q_m is dissipative in the sense of section 2 (see [2], [3]). In fact, with $F = f(Y_0)$ where $f \in L^1(m)$ is strictly positive, the proof of (2.7) yields the existence of a sequence (S_n) of stationary $(\mathcal{G}_{t+})$-stopping times such that

$$\alpha < S_n < \beta \quad \text{on} \quad \{S_n < +\infty\}; \; S_n \downarrow \alpha \quad \text{as} \quad n \to \infty, \, Q_m\text{-a.s.} \tag{3.1}$$

Using the recipe (2.9) we obtain a stationary time S^* with $\alpha < S^* < \beta$ on $\{S^* < +\infty\}$, and $Q_m(S^* \notin \mathbf{R}) = 0$. Define $\mathbf{P} = \mathbf{P}_m$ by (2.11).

(3.2) Proposition. $\mathbf{P}_m$ *is Weil's quasi-process for the pair* $\{m, P_t\}$.

The proof of (3.2) requires some notation. Let $\Omega = \{w \in W : \alpha(w) = 0,\ Y_{\alpha+}(w) \text{ exists in} E\} \cup \{[\Delta]\}$, $\mathcal{F} = \mathcal{G}\mid_\Omega$, $\mathcal{F}_t = \mathcal{G}_t \mid_\Omega$, $X_t = Y_{t+} \mid_\Omega \quad (t \geq 0)$. Define shifts $\theta_t : W \to W$ by

$$\begin{aligned} (\theta_t w)(s) &= w(t+s), & s > 0,\ t \in \mathbf{R}, \\ &= \Delta, & s \leq 0,\ t \in \mathbf{R}, \end{aligned}$$

and note that $\theta_t(\Omega) = \Omega$, $\theta_t(\{\alpha < t\}) = \Omega$. For $x \in E$ let P^x denote the probability measure on $(\Omega, \mathcal{F})$ under which $(X_t : t \geq 0)$ is Markovian with semigroup (P_t). If T is a $(\mathcal{G}_{t+})$-stopping time, then for any $F \in \mathcal{F}^+$,

$$(3.3) \qquad Q_m(F \circ \theta_T \mid \mathcal{G}_{T+}) = P^{Y_T}(F) \qquad \text{on} \qquad \{\alpha < T < \beta\}, Q_m\text{-a.s.}$$

See [3].

Proof of (3.2). For (1.2) take $F = f(Y_0)$ ($f \in \mathcal{E}^+$) in (2.12):

$$m(f) = Q_m(f \circ Y_0) = \mathbf{P}_m\left(\int_{\mathbf{R}} f \circ Y_t \, dt\right).$$

Next, let S be any intrinsic stopping time. Fix $t > 0$ and $F \in \mathcal{F}^+$. Then using (2.5) and $Q_m(S^* \notin \mathbf{R}) = 0$ we have

$$\begin{aligned} &\mathbf{P}(f \circ Y_{S+t} F \circ \theta_{S+t};\ S \in \mathbf{R}) \\ (3.4) \qquad &= Q_m(f \circ Y_{S+t}\, F \circ \theta_{S+t};\ S \in \mathbf{R},\ S^* \in]0,1]) \\ &= Q_m(f \circ Y_{S+t}\, F \circ \theta_{S+t};\ S \in]0,1]). \end{aligned}$$

Using (3.3) with $T = S + t$, the last line in (3.4) can be written as

$$Q_m(f \circ Y_{S+t}\, P^{Y_{S+t}}(F);\ S \in]0,1]),$$

and upon reversing steps in (3.4) the last displayed line becomes

$$(3.5) \qquad \mathbf{P}_m(f \circ Y_{S+t}\, P^{Y_{S+t}}(F);\ S \in \mathbf{R}).$$

Comparing (3.5) with the first line in (3.4) we see that (1.3) holds, and (3.2) is proved. ∎

4. Applications

The energy functional L of Meyer is defined for each pair (m, h) = (excessive function, excessive measure) by

$$L(m, h) = \sup\{\mu(h) : \mu U \leq m\}.$$

Here μ denotes the generic measure on E, while $U = \int_0^\infty P_t dt$. Since m is dissipative there is a sequence (μ_n) of measures such that $\mu_n U \uparrow m$; see [3, §4]. For any such sequence we have

$$(4.1) \qquad L(m,h) = \uparrow \lim_n \mu_n(h), \qquad h \quad \text{excessive}$$

as is easily checked. See [5] for a detailed discussion of L.

Using (4.1) and the result of section 3, we shall outline a systematic approach to various identities involving $\mathbf{P}_m$ found by Weil [10,11], Silverstein [9], and others.

Recall the sequence (S_n) of stationary stopping times in (3.1). Define measures μ_n by

$$(4.2) \qquad \mu_n(B) = \mathbf{P}_m(Y_{S_n} \in B;\ S_n \in \mathbf{R}), \qquad B \in \mathcal{E}.$$

Using (3.3) as in the proof of (3.2) we find that

$$(4.3) \qquad \mu_n U f = \mathbf{P}_m\left(\int_{S_n}^\infty f \circ Y_u \, du;\ \ S_n \in \mathbf{R}\right).$$

But $S_n \downarrow \alpha$ as $n \to \infty$ Q_m-a.s.; consulting (1.2) we see that $\mu_n U \uparrow m$. Thus (4.1) holds for the sequence (μ_n) defined by (4.2).

Let H be a positive function on Ω, measurable over the usual Markovian completion of $\sigma\{X_t : t \geq 0\}$, and excessive: $s \to H \circ \theta_s$ is decreasing and right continuous on $[0, +\infty[$. Then $h(x) = P^x(H)$ is excessive for (P_t). Extend H to all of W by setting

$$H^*(w) = \uparrow \lim_{t \downarrow \alpha(w)} H(\theta_t w).$$

Evidently H^* is (σ_t)-invariant, and

$$(4.4) \qquad \mathbf{P}_m(H^*) = L(m,h).$$

To see (4.4) use (4.1) to let $n \to \infty$ in the identity

$$\mathbf{P}_m(H \circ \theta_{S_n}; S_n \in \mathbf{R}) = \mathbf{P}_m(h \circ Y_{S_n}; S_n \in \mathbf{R}) = \mu_n(h),$$

which can be demonstrated by the argument used to prove (3.2).

Weil's energy formula [11] is a special case of (4.4). Let (A_t) be a finite additive functional of X and let $\kappa(dt)$ be the homogeneous random measure on $\mathbf{R}$ extending dA_t. κ is determined for $w \in W$ by

$$\kappa(w, dt) = d_t A_{t-s}(\theta_s w) \qquad \text{if} \qquad \alpha(w) < s < t.$$

If $H = \int_0^\infty f \circ X_t dA_t$, then $H^* = \int_{\mathbf{R}} f \circ Y_t \kappa(dt)$ and (4.4) becomes

$$\mathbf{P}_m\left(\int_{\mathbf{R}} f \circ Y_t \kappa(dt)\right) = L(m, U_A f),$$

where U_A is the potential operator for (A_t). Also, if $\varphi: [0, +\infty] \to [0, +\infty]$ is increasing then $H \equiv \varphi(A_\infty)$ is excessive, and

$$\mathbf{P}_m(\varphi(\kappa(\mathbf{R}))) = L(m, P^{\cdot}(\varphi(A_\infty))). \tag{4.5}$$

The case $\varphi(x) = x^2/2$ in (4.5) is Weil's energy formula.

Silverstein [9] has obtained other cases of (4.4) under duality hypotheses. We consider one example, that of hitting times and capacities. Let T_B denote the hitting time of the Borel set $B \subset E$ by the process X. The excessive function $P_B 1 \equiv P^{\cdot}(T_B < \infty)$ corresponds to the excessive random variable $H = 1_{\{T_B < \infty\}}$. If we let $\rho(w) = \{Y_t(w): \alpha(w) < t < \beta(w)\}$ denote the range of Y, then $H^* = 1_{\{\rho \cap B \neq \phi\}}$. By (4.4),

$$\Gamma(B) \equiv L(m, P_B 1) = \mathbf{P}_m(\rho \cap B \neq \phi). \tag{4.6}$$

The set function Γ is discussed in [5] where it is shown to be a natural extension of both the capacity C and the cocapacity $\hat{C}$, which we now recall.

Following [4,5], for $B \in \mathcal{E}$ define stationary times $\tau_B = \inf\{t: Y_t \in B\}$, $\lambda_B = \sup\{t: Y_t \in B\}$, and set

$$\hat{C}(B) = Q_m(0 < \tau_B \leq 1), \quad C(B) = Q_m(0 < \lambda_B \leq 1).$$

A set B is transient if $\lambda_B < +\infty$ Q_m-a.s. in which case $C(B)$ is the capacity of B. Similarly $\hat{C}(B)$ is the cocapacity of B if $\tau_B > -\infty$ Q_m-a.s. (i.e., if B is cotransient). By (2.5),

$$\hat{C}(B) = \mathbf{P}_m(\tau_B \in \mathbf{R}), \quad C(B) = \mathbf{P}_m(\lambda_B \in \mathbf{R}).$$

But $\{\rho \cap B \neq \phi\} = \{\tau_B < +\infty\} = \{\lambda_B > -\infty\}$, and so

$$\begin{aligned} \Gamma(B) &= C(B) + \mathbf{P}_m(\lambda_B = +\infty) \\ &= \hat{C}(B) + \mathbf{P}_m(\tau_B = -\infty). \end{aligned} \tag{4.7}$$

thus $\Gamma(B) = C(B)$ if B is transient, and $\Gamma(B) = \hat{C}(B)$ if B is cotransient. See [4,5] for further details.

The reader can check that the familiar properties (monotonicity, strong subadditivity, etc.) of Γ, C, $\hat{C}$ follow easily from (4.6) and (4.7). Consider for example the "inclusion-exclusion" inequality

$$\Gamma(B_1 \cap \cdots \cap B_n) \leq \sum (-1)^{k+1} \Gamma(B_{i_1} \cup \cdots \cup B_{i_k}), \tag{4.8}$$

where the sum extends over all nonempty subsets $\{i_1, \ldots, i_k\}$ of $\{1, 2, \ldots, n\}$, and $n \in \mathbf{N}$ is arbitrary. (Since $\Gamma(B)$ may be $+\infty$ for certain B's, for a precise statement of (4.8) one should move the terms with k even to the L.H.S. of the inequality, dropping the sign in the process.) For a proof let $J(B)$ denote the indicator of $\{\rho \cap B \neq \phi\}$ and take $\mathbf{P}_m$ expectations in the inequality

$$J(B_1 \cap \cdots \cap B_n) \leq \prod_1^n J(B_i) = \sum (-1)^{k+1} J(B_{i_1} \cup \cdots \cup B_{i_k}).$$

It can be shown that (4.8) remains valid if Γ is replaced by C (resp. $\hat{C}$) provided at least one of the B_i's is transient (resp. cotransient). Note also that (4.8) (for all $n \in \mathbf{N}$) coupled with the monotonicity of Γ is equivalent to the fact that Γ is "alternating of order ∞" in the sense of Choquet.

References

1. B. W. ATKINSON and J. B. MITRO. Applications of Revuz and Palm type measures for additive functionals in weak duality. Seminar on Stochastic Processes, 1982, pp. 23–49. Birkhäuser, Boston-Basel-Stuttgart, 1983.
2. E. B. DYNKIN. An application of flows to time shift and time reversal in stochastic processes. Trans. Am. Math. Soc. **287** (1985) 613–619.
3. P. J. FITZSIMMONS and B. MAISONNEUVE. Excessive measures and Markov processes with random birth and death. Prob. Th. Rel. Fields **72** (1986) 319–336.
4. R. K. GETOOR and J. STEFFENS. Capacity theory without duality. Prob. Th. Rel. Fields **73** (1986) 415–445.
5. R. K. GETOOR and J. STEFFENS. The energy functional, balayage, and capacity. To appear in Annales de l'Inst. H. Poincaré.
6. G. A. HUNT. Markov chains and Martin boundaries. Ill. J. Math. **4** (1960) 313–340.
7. S. E. KUZNETSOV. Construction of Markov processes with random times of birth and death. Th. Prob. App. **18** (1974) 571–574.
8. J. NEVEU. Sur les mesures de Palm de deux processus ponctuels stationnaires. Z. Wahrsch. verw. Geb. **34** (1976) 199–203.
9. M. L. SILVERSTEIN. Applications of the sector condition to the classification of subMarkovian semigroups. Trans. Am. Math. Soc. **244** (1978) 103–146.

10. M. WEIL. Quasi-processus. Séminaire de Probabiités IV, pp. 216–239. Lecture Notes in Math. **124**. Springer-Verlag, Berlin-Heidelberg-New York, 1970.

11. M. WEIL. Quasi-processus et énergie, Séminaire de Probabilités V, pp. 347–361. Lecture Notes in Math. **191**. Springer-Verlag, Berlin-Heidelberg-New York, 1971.

P. J. FITZSIMMONS
Department of Mathematics
University of California, San Diego
La Jolla, California 92093

MORE ABOUT CAPACITY AND EXCESSIVE MEASURES

by

R. K. Getoor* and J. Steffens

1. Introduction

In [14] we introduced the set function

$$(1.1) \qquad \Gamma(B) = \Gamma_{m,u}(B) = L(m, P_B u) = L(R_B m, u)$$

as a "good" capacity. In this paper we develop some additional properties of $\Gamma(B)$. We are especially interested in obtaining expressions for $\Gamma(B)$ as a supremum that are analogous to the classical result of de la Vallée-Poussin which states that the (Newtonian) capacity of a Borel set B is the supremum of $\mu(1)$ over all measures μ with compact support in B and whose potential is bounded by 1. Results of this character are contained in sections 3 and 4. In section 5 we obtain some expressions for Γ as an infimum. These results are reminiscent of the work of Fuglede [19] and are in some sense dual to the classical result. Section 6 contains additional formulas for $\Gamma^q(B)$—the q-capacity of B; that is, the capacity relative to the q-subprocess. Of special interest is (6.7) which shows that $q \to \Gamma^q(B)$ is a subordinator exponent (i.e., has a completely monotone derivative) provided it is finite for one value of $q > 0$. Section 2 contains two results in the potential theory of excessive measures that are needed in section

* Research supported by NSF Grant DMS-8419377.

3. But these are of considerable interest in their own right. Theorem 2.18 sheds light on an old problem going back to [2]. See also [17] and [15].

We assume the reader is familiar with [14] and the present paper may be viewed as a continuation of [14]. In particular we use the notation and terminology of [14] without special mention. However, we emphasize that the underlying process X is a Borel right process as in [14]. Because of the vagaries of printers, word processers, and typewriters we record the fact that $\mathbf{E}^q$ denotes the class of q-excessive functions for X, and, as usual, $\mathbf{E} = \mathbf{E}^0$.

In [7] Fitzsimmons has obtained an extremely nice probabilistic expression for Γ which we shall describe. Suppose $m \in \text{Dis}$—really the only interesting case—then Fitzsimmons shows that there exists $S: W \to [-\infty, \infty]$ that is stationary in the sense that $S \circ \theta_t = S - t$ and satisfies $Q_m^u(S \notin \mathbf{R}) = 0$. Then

$$\Gamma(B) = Q_m^u(\tau_B < \infty;\ 0 < S < 1) = Q_m^u(\lambda_B > -\infty;\ 0 < S < 1), \tag{1.2}$$

where $\tau_B = \inf\{t: Y_t \in B\}$ and $\lambda_B = \sup\{t: Y_t \in B\}$ as in [14]. In particular these expressions do not depend on the choice of S subject to the above conditions. Note that

$$\{\tau_B < \infty\} = \{\lambda_B > -\infty\} = \{Y_t \in B \quad \text{for some} \quad t\}.$$

From (1.2) one may read off many of the properties of Γ. For example, Γ is alternating of order infinity.

We would like to thank P. J. Fitzsimmons for several very helpful conversations, especially in connection with sections 5 and 6.

2. Two Results for Excessive Measures

In this section we shall formulate and prove two results for excessive measures that are analogous to well-known facts about excessive functions. For the first of these let $f \in p\mathcal{E}^*$ and vanish off a nearly Borel set B. Then $P_B U f = Uf$, and consequently if $u \in \mathbf{E}$ with $Uf \leq u$, then $Uf \leq P_B u$. (Of course, it suffices to suppose that $Uf \leq u$ on B.) Since R_B is the analog for excessive measures of

P_B for excessive functions, one might conjecture that if $m \in \text{Exc}$ and $\mu U \leq m$ with μ carried by a Borel set B, then $\mu U \leq R_B m$. However, simple examples show that this is false. (Consider translation to the right on $\mathbf{R}$ at unit speed and take $m = \epsilon_0 U$, $\mu = \epsilon_0$ and $B = \{0\}$.)

In order to formulate the correct result we must introduce the subset W_q of W defined in (5.7) of [13] or below (3.5) of [9]. To this end fix $m \in \text{Exc}$ and choose $q \in \mathcal{E}$ with $q > 0$ and $m(q) < \infty$. Let D be a countable, uniformly dense subset of the bounded real valued uniformly continuous (relative to some metric on E compatible with the topology of E) functions on E. Then W_q is defined by the conditions: (i) $\alpha > -\infty$, (ii) $Y_{\alpha+}$ exists in E, (iii) $U_r g(Y_{\alpha+1/n}) \to U_r g(Y_{\alpha+})$ as $n \to \infty$ for $g \in D$ and r a strictly positive rational, and (iv) $Uq(Y_{\alpha+1/n}) \to Uq(Y_{\alpha+})$. Note that W_q depends on m and q only through condition (iv). Clearly $W_q \in \mathcal{G}^0_{\alpha+}$ and is invariant, $\theta_t^{-1} W_q = W_q$ for $t \in \mathbf{R}$. Define

$$
\begin{aligned}
(2.1) \qquad & Y_t^*(w) = Y_t(w) \quad \text{if} \quad t \neq \alpha(w) \quad \text{or} \quad t = \alpha(w) \quad \text{and} \quad w \notin W_q \\
& Y_\alpha^*(w) = Y_{\alpha+}(w) \quad \text{if} \quad w \in W_q.
\end{aligned}
$$

This extended process Y^* differs slightly from the process $\overline{Y}$ introduced in (3.8) of [6]. In some sense Y^* is the maximal extension of Y for which the strong Markov property holds at α when $\alpha > -\infty$. To formulate this more precisely define for each $t \in \mathbf{R}$, $\gamma_t^* : W \to \Omega$ by

$$
\begin{aligned}
(2.2) \quad \gamma_t^* w(s) &= w(t+s) && \text{for } s \geq 0 && \text{if } \alpha(w) < t \\
&= w(t+s+) && \text{for } s \geq 0 && \text{if } \alpha(w) = t \text{ and } w \in W_q \\
&= \Delta && \text{for } s \geq 0 && \text{otherwise.}
\end{aligned}
$$

Note that (b_t is the birthing operator—see (5.9) of [14])

$$
(2.3) \qquad \{\alpha < t\} \cup (\{\alpha = t\} \cap W_q) = b_t^{-1} W_q.
$$

With the usual abuse of notation we shall write $W_q \circ b_t$ for $b_t^{-1} W_q$. It is immediate that $X_s \circ \gamma_t^* = Y_{s+t}^*$ for $s \geq 0$ on $W_q \circ b_t$. The following result is a simple variant of (5.8) in [13], (3.3) in [9], or (3.10) in [6] and, as such, we shall omit the proof.

(2.4) Theorem. *Let T be a $(\mathcal{G}_t^m)$ stopping time with $T \geq \alpha$, $F \in p\mathcal{F}^*$, and $\Lambda \in \mathcal{G}_T^m$. Then*

$$Q_m[F \circ \gamma_T^*;\ \Lambda;\ W_q \circ b_T;\ T \in \mathbf{R}] = Q_m[P^{Y^*(T)}(F);\ \Lambda;\ W_q \circ b_T\ \ T \in \mathbf{R}].$$

If $B \in \mathcal{E}$, define

$$\tau_B = \inf\{t: Y_t \in B\};\quad \tau_B^* = \inf\{t: Y_t^* \in B\}, \tag{2.5}$$

where the infimum of the empty set is $+\infty$. Then both τ_B and τ_B^* are $(\mathcal{G}_{t+}^*)$ stopping times and are *stationary* in the sense that $\tau \circ \theta_t = \tau - t$. Moreover $\alpha \leq \tau_B^* \leq \tau_B$. We write R_B^* and R_B for the balayage operations defined in [9] using τ_B^* and τ_B; that is,

$$R_B^*\xi(f) = Q_\xi[f \circ Y_0; \tau_B^* < 0] = Q_\xi[f \circ Y_0^*; \tau_B^* < 0] \tag{2.6}$$

for $\xi \in \mathrm{Exc}$ while R_B, defined similarly, is the usual balayage operator. See [9] or [14]. It is immediate that $R_B\xi \leq R_B^*\xi$. Finally let

$$m = \nu U + \rho \tag{2.7}$$

be the Riesz decomposition of m into a potential νU and a harmonic excessive measure ρ. See (3.7) of [9] or [12]. In particular it is shown in [9] that

$$Q_{\nu U}(F) = Q_m(F; W_q),\quad Q_\rho(F) = Q_m(F; W_q^c), \tag{2.8}$$

for $F \in p\mathcal{G}^*$ and that ν is given by

$$\nu(f) = Q_m(f \circ Y_{\alpha+};\ 0 < \alpha < 1;\ W_q) \tag{2.9}$$

for $f \in p\mathcal{E}$.

We now are prepared to state our first result.

(2.10) Theorem. *Suppose $\mu U \leq m$ and μ is carried by B. Then $\mu U \leq R_B^* m$. Moreover $Q_m(\tau_B^* < \tau_B) = 0$ if and only if $\nu(B - B^r) = 0$, and in this case $\mu U \leq R_B m$. Here B^r is the set of regular points of B.*

Proof. Let $\xi = \mu U \leq m$. The proof is based on a result of Fitzsimmons in [6]. See also [8]. According to (6.36) and the sentence below (6.46) in [6], there exists an increasing family $\{T(u);\ 0 \leq u \leq 1\}$ of stationary $(\mathcal{G}_t^m)$ stopping times with $\alpha \leq T(u) < \beta$ if $T(u) < \infty$ (i.e., intrinsic times as defined in [9]) such that

$$\xi = \int_0^1 R_{T(u)} m \, du. \tag{2.11}$$

Therefore using (5.3) of [9] we have

$$Q_\xi(F) = \int_0^1 Q_m[F \circ b_{T(u)};\ T(u) < \infty] du. \tag{2.12}$$

In addition from (3.3) in [9]—see (2.9)—

$$\mu(f) = Q_\xi[f \circ Y_{\alpha+};\ 0 < \alpha < 1, W_q]. \tag{2.13}$$

If $t \geq \alpha$, $\alpha \circ b_t = \alpha \vee t = t$ and $Y_{\alpha+} \circ b_t = Y_{t+}$ and so combining (2.12) and (2.13) gives

$$\mu(f) = \int_0^1 du Q_m[f \circ Y_{T(u)+};\quad 0 < T(u) < 1;\quad W_q \circ b_{T(u)}]. \tag{2.14}$$

Next define

$$\begin{aligned} S(u) &= T(u) \quad \text{on} \quad \Gamma = \{Y_{T(u)+} \in B;\ W_q \circ b_{T(u)}\} \\ &= +\infty \quad \text{otherwise.} \end{aligned} \tag{2.15}$$

It is readily checked that $\Gamma \circ \theta_t = \Gamma$ for $t \in \mathbf{R}$ and $\Gamma \in \mathcal{G}_{T(u)}^m$. Consequently each $S(u)$ is a stationary $(\mathcal{G}_t^m)$ stopping time, and since μ is carried by B it follows from (2.14) that

$$\mu(f) = \int_0^1 du\, Q_m[f \circ Y_{S(u)}^*;\ 0 < S(u) < 1;\ W_q \circ b_{S(u)}]. \tag{2.16}$$

We now compute $\mu U(f)$ using (2.16) and (2.4) for the second equality below. Also let $H(u) = W_q \circ b_{S(u)}$ and note that $H(u) \circ \theta_t = H(u)$. Then

$$\begin{aligned} \mu[Uf] &= \int_0^1 du\, Q_m[Uf \circ Y_{S(u)}^*;\ 0 < S(u) < 1;\ H(u)] \\ &= \int_0^1 du\, Q_m \left[\int_0^\infty f \circ Y_{S(u)+t}^* \, dt;\ 0 < S(u) < 1;\ H(u) \right] \\ &= \int_0^1 du \int_{-\infty}^\infty dt\, Q_m[f \circ Y_t; S(u) < t;\ 0 < S(u) < 1;\ H(u)] \end{aligned}$$

$$= \int_0^1 du \int_{-\infty}^{\infty} dt\ Q_m[f \circ Y_0;\ S(u) < 0;\ -t < S(u) < -t+1;\ H(u)]$$

$$= \int_0^1 du\ Q_m[f \circ Y_0;\ S(u) < 0;\ H(u)].$$

But $Y^*_{S(u)} \in B$ on $\{S(u) < \infty\}$ and hence $\{S(u) < 0\} \subset \{\tau^*_B < 0\}$. Consequently

$$\mu U(f) \leq Q_m[f \circ Y_0;\ \tau^*_B < 0] = R^*_B m(f),$$

proving the first assertion in (2.10).

It remains to show $Q_m(\tau^*_B < \tau_B) = 0$ if and only if $\nu(B - B^r) = 0$. But

$$\{\tau^*_B < \tau_B\} = \{Y_{\alpha+} \in B;\ \alpha < \tau_B\} \cap W_q \subset W_q$$

and so $Q_\rho(\tau^*_B < \tau_B) = 0$ by (2.8). Since $Q_{\nu U}$ is carried by W_q, if νP denotes the entrance law $(\nu P_t)_{t>0}$ one has

$$\begin{aligned} Q_{\nu U}(\tau^*_B < \tau_B) &= Q_{\nu U}(Y_{\alpha+} \in B, \alpha < \tau_B) \\ &= \int_{-\infty}^{\infty} dt\ Q_{\nu P}(Y_{\alpha+} \in B, \alpha < \tau_B) \\ &= \int_{-\infty}^{\infty} dt\ P^\nu(X_0 \in B, 0 < T_B) \end{aligned}$$

where for the second equality we have used the invariance of $\{Y_{\alpha+} \in B, \alpha < \tau_B\}$ and the third follows because the law of $(Y^*_t)_{t\geq 0}$ under $Q_{\nu P}$ is the same as the law of $(X_t)_{t\geq 0}$ under P^ν. See (3.3) of [9], for example. But $P^\nu(X_0 \in B, 0 < T_B) = 0$ if and only if $\nu(B - B^r) = 0$. Obviously $R_B m = R^*_B m$ if $Q_m(\tau^*_B < \tau_B) = 0$ and so $\mu U \leq R_B m$ in this case. ∎

(2.17) **Remark.** If $f > 0$ with $m(f) < \infty$, then

$$R^*_B m(f) - R_B m(f) = Q_m[f \circ Y_0;\ \tau^*_B < 0 < \tau_B],$$

and so it follows readily that $R_B m = R^*_B m$ if and only if $\nu(B - B^r) = 0$.

We turn now to our second result. In [2] it was shown that if X and $\hat{X}$ are standard processes in strong duality with respect to m, if Uf is lower semicontinuous for positive continuous f with compact support, and if X is transient, then μ does not charge polar sets whenever $U\mu$ is bounded. These hypotheses were relaxed somewhat in [17]. However, Glover constructed a very simple example in [15] of two standard processes in strong duality and a measure μ carried by a polar set with $U\mu$ bounded. Glover's example is described in the first paragraph of section 3. If we drop the duality assumption one might expect that $\mu U \leq km$ implies that μ does not charge m-polar sets under some hypotheses. Actually we can do slightly better. Fitzsimmons in [6] has introduced a weak order between excessive measures which he writes $\xi \leftarrow m$. According to (6.41) of [6], $\xi \leftarrow m$ if and only if $\xi = \sum \xi_n$ with $\xi_n \in$ Exc and $\xi_n \leq m$ for each n. In particular if $\xi \leq km$, then $\xi \leftarrow m$. In the next theorem $m \in$ Exc and $m = \nu U + \rho$ is its Riesz decomposition as in (2.7).

(2.18) Theorem. *Suppose $\mu U \leftarrow m$. Then $\mu(B) = 0$ whenever B is m-polar and $\nu(B) = 0$.*

Proof. Let $\mu U = \sum \xi_n$ with $\xi_n \leq m$. Since $\xi_n \leq \mu U$, one has $\xi_n = \mu_n U$—see [11] or [5]—and since $\mu U = \sum \mu_n U$ the uniqueness theorem for potentials gives $\mu = \sum \mu_n$. Therefore it suffices to prove (2.18) when $\mu U \leq m$. Then μ is given by (2.14). From (2.3),

$$W_q \circ b_{T(u)} = \{\alpha < T(u)\} \cup (\{\alpha = T(u)\} \cap W_q),$$

and so we may write $\mu = \mu_1 + \mu_2$ where in (2.14), μ_1 corresponds to integrating in w over $\{\alpha < T(u)\}$ and μ_2 corresponds to integrating over $\{\alpha = T(u)\} \cap W_q$. Consequently

$$\mu_1(f) = \int_0^1 du\, Q_m[f \circ Y_{T(u)};\ 0 < T(u) < 1,\ \alpha < T(u)], \tag{2.19}$$

and so if B is m-polar, then $\mu_1(B) = 0$. Hence μ_1 does not charge m-polars. Next

$$\mu_2(f) = \int_0^1 du\, Q_m[f \circ Y_{\alpha+};\ 0 < \alpha = T(u) < 1;\ W_q].$$

It follows from this, (2.8), and (2.9) that

$$\mu_2(f) \leq Q_{\nu U}[f \circ Y_{\alpha+};\ 0 < \alpha < 1;\ W_q] = \nu(f).$$

Therefore $\mu_2(B) = 0$ whenever $\nu(B) = 0$. Combining these statements gives (2.18). ∎

(2.20) **Remarks.** Note that the proof of (2.18) actually shows that if $\mu U \leftarrow m$, then μ doesn't charge $m - Y^*$ polar sets; that is, sets $B \in \mathcal{E}$ with $Q_m(\tau_B^* < \infty) = 0$. No doubt this is the proper formulation of (2.18).

3. Capacity as a Supremum I

In classical potential theory the capacity of a set B may be expressed as the supremum of $\mu(1)$ as μ ranges over all finite measures with compact support contained in B and $U\mu \leq 1$. This extends to the situation of standard processes in strong duality with respect to m subject to some regularity conditions. See, for example, the discussion on page 286 of [1]. However, this is not true in general even for standard processes in strong duality. Consider Glover's example in [15]. Here E is the unit circle in $\mathbf{R}^2$ and x_0 a fixed point of E. Let X (resp. $\hat{X}$) be translation at unit speed counterclockwise (resp. clockwise) around E killed at the time of hitting x_0. Let m be Haar measure on E. Then X and $\hat{X}$ are in strong duality with respect to m. Since $K = \{x_0\}$ is polar, it must have capacity zero. But, as is easily checked, $U\epsilon_{x_0} \leq 1$. Therefore one can not expect the classical results to hold even for standard processes in strong duality.

In the remainder of this section we fix $m \in \text{Dis}$ and $u \in \mathbf{E}$ with $m(u = \infty) = 0$. As in (1.1)

$$\Gamma(B) = L(m, P_B u) = L(R_B m, u) \tag{3.1}$$

is the capacity of $B \in \mathcal{E}$. We shall obtain some expressions for $\Gamma(B)$ as a supremum that are generalizations of the classical results described above, but one should always keep Glover's example in mind.

We shall need the following result that is Theorem XII-38 in [4]. In its statement we use the same notation as in [4].

(3.2) Lemma. *Let $\lambda \in$ Dis. Then there exists a sequence (η_n) of finite measures with $\eta_n U \uparrow \lambda$. One may suppose that $\eta_n \leq n\lambda$ for each n.*

(3.3) **Remarks.** In [4] this is proved assuming that U is proper so Exc = Dis. But choosing $g > 0$ with $\lambda(g) < \infty$ one has $Ug < \infty$ a.e. λ because $\lambda \in$ Dis. By restricting X to the absorbing set $\{Ug < \infty\}$ the result extends to $\lambda \in$ Dis for general X. See [9], [13], or [14] for similar reductions. The η_n constructed in [4] are of the form $\eta_n = n[\lambda_n - n\lambda_n U^n]$ where each $\lambda_n \in$ Pur and $\lambda_n \leq \lambda$. Hence $\eta_n \leq n\lambda_n \leq n\lambda$.

Let $\mathcal{M}_K$ denote the class of finite measures on $(E, \mathcal{E})$ with compact support. Supp(μ) will denote the support of a measure μ on E. The letter B with or without subscripts will denote a Borel subset of E. Let $m = \nu U + \rho$ be the decomposition of m into its potential and harmonic parts as in (2.7). Finally we remind the reader that m *is assumed to be dissipative throughout this section.*

(3.4) Lemma. *(i) If $\mu U \leq R_B m$, then $\mu(u) \leq \Gamma(B)$. (ii) If $\mu U \leq m$, then $\mu(P_B u) \leq \Gamma(B)$. (iii) If $\mu U \leq m$ and either (a) μ is carried by B^r or (b) μ is carried by B and $\nu(B - B^r) = 0$, then $\mu(u) \leq \Gamma(B)$.*

Proof. If $\mu U \leq R_B m$, then

$$\mu(u) = L(\mu U, u) \leq L(R_B m, u) = \Gamma(B)$$

proving (i). If $\mu U \leq m$, then $\mu P_B U = R_B(\mu U) \leq R_B m$ by (5.14) of [9]. Hence

$$\mu(P_B u) = L(\mu U, P_B u) = L(R_B(\mu U), u) \leq L(R_B m, u) = \Gamma(B),$$

proving (ii). If $\mu U \leq m$ and μ is carried by B^r, then $\mu(P_B u) = (\mu P_B)(u) = \mu(u)$, and so (iii-a) is a consequence of (ii). If μ is carried by B and $\nu(B - B^r) = 0$, then according to Theorem 2.10, $\mu U \leq R_B m$. Thus (iii-b) follows from (i). ∎

(3.5) Proposition. *Given $B \in \mathcal{E}$ there exists an increasing sequence (K_n) of compact subsets of B with $R_{K_n} m \uparrow R_B m$ and $\Gamma(K_n) \uparrow \Gamma(B)$.*

Proof. According to the remarks following (12.15) in [10] there exists an increasing sequence (K_n) of compact subsets of B with $T_{K_n} \downarrow T_B$ a.s. P^m. It is easily checked that $\tau_{K_n} \downarrow \tau_B$ a.s. Q_m and hence $R_{K_n}m \uparrow R_Bm$. Therefore using the properties of L—see e.g. section 3 of [14]—

$$\Gamma(K_n) = L(R_{K_n}m, u) \uparrow L(R_Bm, u) = \Gamma(B).$$ ∎

(3.6) Proposition. *Given $B \in \mathcal{E}$ there exists a sequence $(\mu_n) \subset \mathcal{M}_K$ with the following properties: (i)* $\mathrm{Supp}(\mu_n) \subset B$. *(ii) μ_n does not charge m polar sets. (iii) $\mu_n U \uparrow R_Bm$. (iv) $\mu_n(u) \uparrow \Gamma(B)$.*

Proof. Let (K_n) be chosen as in (3.5). By (3.2) there exist finite measures η_k with $\eta_k \le km$ and $\eta_k U \uparrow m$. Then $\eta_k P_{K_n} U \uparrow R_{K_n}m$ and $R_{K_n}m \uparrow R_Bm$. Since $\eta_k P_{K_n} U$ increases with both n and k, defining $\mu_n = \eta_n P_{K_n}$ one obtains $\mu_n U \uparrow R_Bm$. Clearly $\mathrm{Supp}(\mu_n) \subset K_n \subset B$ and μ_n is finite for each n. Moreover

$$\mu_n(u) = L(\mu_n U, u) \uparrow L(R_Bm, u) = \Gamma(B).$$

It remains to verify (ii). If A is m-polar, then $m(A) = 0$ because m is excessive. Therefore $\mu_n(A) = \eta_n P_{K_n}(A) \le n \cdot m P_{K_n}(A) = 0$. ∎

(3.7) Theorem.

(i) $\Gamma(B) = \sup\{\Gamma(K): K \subset B,\ K \text{ compact}\}$.

(ii) $\Gamma(B) = \sup\{\mu(u): \mu U \le R_Bm\}$.

(iii) $\Gamma(B) = \sup\{\mu(u): \mu \in \mathcal{M}_K, \mathrm{Supp}(u) \subset B, \mu U \le R_Bm\}$.

(iv) If $\nu(B - B^r) = 0$, then

$$\Gamma(B) = \sup\{\mu(u): \mu \in \mathcal{M}_K,\ \mathrm{Supp}\,\mu \subset B, \mu U \le m\}.$$

(v) If $B - B^r$ is m-polar, then

$$\Gamma(B) = \sup\{\mu(u): \mu \in \mathcal{M}_K^m,\ \mathrm{Supp}(\mu) \subset B,\ \mu U \le m\},$$

where $\mathcal{M}_K^m$ is the set of those μ in $\mathcal{M}_K$ which do not charge m-polar sets.

Proof. (i) is immediate from (3.5), while (ii) and (iii) follow from (3.4-i) and (3.6). (iv) follows from (iii) and (3.4-iii-b). If $B-B^r$ is m polar and μ doesn't charge m-polars then μ is carried by B^r, and so (v) follows from (3.4-iii-a) and (3.6). ∎

(3.8) **Remarks.** In view of (3.6) one may replace $\mathcal{M}_K$ by $\mathcal{M}_K^m$ in (iii) and (iv) of (3.7). If ν does not charge semipolar sets, then (3.7-iv) reduces to the familiar statement in classical potential theory. Since $\Gamma_{m,u} = \Gamma_{\nu U,u} + \Gamma_{\rho,u}$ and $\Gamma_{\nu U,u}(B) = L(\nu U, P_B u) = \nu(P_B u)$ one obtains the following

$$(3.9) \qquad \Gamma(B) = \nu(P_B u) + \sup\{\mu(u) : \mu \in \mathcal{M}_K,\ \mathrm{Supp}(\mu) \subset B,\ \mu U \le \rho\}$$

by using (3.7-iv) for $\Gamma_{\rho,u}$. Again one may replace $\mathcal{M}_K$ by $\mathcal{M}_K^m$ or even $\mathcal{M}_K^\rho$ in (3.9).

(3.10) **Examples.** (i). (3.7-v) is not valid for all B. Let X be translation to the right at unit speed on $\mathbf{R}$. Let λ be Lebesgue measure and $m = \epsilon_0 U + \frac{1}{2}\lambda$. Take $u = 1$ and $B = \{0\}$. Now $\epsilon_0 U \le m$ and B is not m-polar. But $L(\epsilon_0 U, P_B 1) = P_B 1(0) = 0$ and $L(\lambda, P_B 1) = \lim_{n\to\infty} \epsilon_{-n} P_B 1 = 1$. Therefore $\Gamma(B) = \frac{1}{2}$ while $\epsilon_0(B) = 1$. This also shows that (3.7-iv) is not valid for general m. (ii) One may not replace m-polar by polar in (3.7-v). Let X be as in example (i) and take $m = \epsilon_0 U$, $B = \{0\}$, and $u = 1$. Then $\{0\}$ is m-polar but not polar, while $\Gamma(B) = 0$ and $\epsilon_0(B) = 1$.

4. Capacity as a Supremum II

In this section we give some results expressing $\Gamma(B)$ as a supremum of $m(f)$ as f ranges over certain classes of functions. These results are, in some sense, dual to those in section 3. They are related to semi-classical potential theory as developed by Kac, Ciesielski, and Stroock. See, for example, [3] and [18]. As in the previous section $m \in \mathrm{Dis}$ and $u \in \mathbf{E}$ are fixed with $m(u = \infty) = 0$. We also follow the notational conventions of §3, and adopt the convention that f with or without subscripts denotes a function in $p\mathcal{E}$. Finally a.e. means a.e. m unless explicitly stated otherwise.

(4.1) Proposition. *(i) If $Uf \leq P_B u$ a.e., then $m(f) \leq \Gamma(B)$. (ii) If $Uf \leq u$ a.e. and f vanishes off B, then $m(f) \leq \Gamma(B)$.*

Proof. If $Uf \leq P_B u$ a.e., then

$$m(f) = L(m, Uf) \leq L(m, P_B u) = \Gamma(B).$$

If f vanishes off B, then $Uf = P_B Uf$. If $Uf \leq u$ a.e., then $\{u < Uf\}$ is finely open and has m measure zero; hence, it is m-polar. Therefore $P^m[u \circ X_{T_B} < Uf \circ X_{T_B}] = 0$. Consequently $P_B Uf \leq P_B u$ a.e., and so (ii) follows from (i). ∎

Since $m \in$ Dis there exists $(f_k) \subset bp\mathcal{E}$ with $Uf_k \uparrow P_B u$ a.e. See, for example, the discussion at the beginning of §3 in [14]. Thus, in view of (4.1-i), one has

$$\Gamma(B) = \sup\,\{m(f) \colon Uf \leq P_B u \text{ a.e.}\}. \tag{4.2}$$

In order to obtain a result in some sense dual to (3.7-iv) we introduce the Lebesgue penetration time S_B of B:

$$S_B = \inf\,\{t \colon \int_0^t 1_B \circ X_s ds > 0\}. \tag{4.3}$$

In view of the a.e. appearing in (4.1) and (4.2) it is not surprising that S_B should enter our discussion. In fact, the semi-classical potential theory of Kac is based on S_B rather than T_B. A famous result of Hunt states that if X is transient and B finely open then there exists $(f_n) \subset pb\mathcal{E}$ with each f_n vanishing off B and $Uf_n \uparrow P_B u$. See [16] or [11]. The argument in these references actually shows that for a general B there exists $(f_n) \subset pb\mathcal{E}$ with each f_n vanishing off B and $Uf_n \uparrow P_{S_B} u$. For general X and $m \in$ Dis choose $g > 0$ with $Ug < \infty$ a.e. Then X restricted to the absorbing set $\{Ug < \infty\}$ is transient, and so there exist $f_n \in pb\mathcal{E}$ vanishing off $A = B \cap \{Ug < \infty\}$ with $Uf_n \uparrow P_{S_A} u$ on $\{Ug < \infty\}$. Since $S_A = S_B$ a.s. P^x if $Ug(x) < \infty$ we see that $Uf_n \uparrow P_{S_B} u$ a.e. Consequently $m(f_n) = L(m, Uf_n) \uparrow L(m, P_{S_B} u)$ and so one has the following result.

(4.4) Proposition. *If $P_B u = P_{S_B} u$ a.e., then*

(4.5) $$\Gamma(B) = \sup\{m(f): Uf \leq u \text{ a.e.};\ f \quad \textit{vanishes off} \quad B\}.$$

(4.6) **Example.** Let X be Brownian motion killed when it leaves $]-1,1[$, $B = \{0\}$, m Lebesgue measure on $]1,1[$, and $u = 1$. Then $B = B^r$. If $f = 0$ off B, then $Uf = 0$ and $m(f) = 0$. But $\Gamma(B) > 0$ so (4.5) fails for general B. Moreover this also shows that one can not restrict f to vanish off B in (4.2).

5. Capacity as an Infimum

In this section we obtain some expressions for capacity as an infimum. It is interesting that in contrast to sections 3 and 4 we obtain better results in taking infima over functions (the analog of section 4) than over measures (the analog of section 3). We also obtain a result expressing $\Gamma(B)$ as an infimum of $\Gamma(G)$ as G ranges over finely open Borel supersets of B under some additional hypotheses. This in turn leads to results that are in the same spirit as those in [19].

Recall that $v \in p\mathcal{E}^*$ is *supermedian* provided $P_t v \leq v$ for each $t \geq 0$ and is *strongly supermedian* provided $P_T v \leq v$ for all stopping times T. Let $\mathbf{S}$ (resp. $\mathbf{S}^s$) denote the class of supermedian (resp. strongly supermedian) functions. If $v \in \mathbf{S}$, then $\hat{v} = \lim_{t \downarrow 0} P_t v$ is the largest excessive function dominated by v and is called the excessive regularization of v. The set $\{\hat{v} < v\}$ is of potential zero and hence has λ measure zero for all $\lambda \in$ Exc. Finally recall that L is extended to Exc $\times\ \mathbf{S}$ by

(5.1) $$L(\lambda, v) = L(\lambda, \hat{v}).$$

As in sections 3 and 4 we fix $m \in$ Dis and $u \in \mathbf{E}$ with $m(u = \infty) = 0$ and let $\Gamma = \Gamma_{m,u}$. The notational conventions of those sections are still in force.

(5.2) Proposition. $\Gamma(B) = \inf\{L(m, v): v \in \mathbf{S}^s;\ v \geq u \quad \textit{on} \quad B\}$.

Proof. We first claim that if $v \in \mathbf{S}^s$ and $v \geq u$ on B, then $v \geq u$ on $B \cup B^r$. As in the proof of (3.5) given $x \in B^r$, there exists an increasing sequence (K_n) of compact subsets of B with $T_{K_n} \downarrow T_B = 0$ a.s. P^x. Then if $k < \infty$

$$v(x) \geq P_{K_n}v(x) \geq P_{K_n}(u \wedge k)(x),$$

because $v \in \mathbf{S}^s$ and $P_{K_n}(x, \cdot)$ is carried by $K_n \subset B$. By the bounded convergence $P_{K_n}(u \wedge k)(x) \to u(x) \wedge k$ as $n \to \infty$, and then letting $k \to \infty$ we obtain $v(x) \geq u(x)$. Hence $v \geq u$ on $B \cup B^r$ as claimed. Since P_B is carried by $B \cup B^r$, $v \geq P_B v \geq P_B u$ and so $\hat{v} \geq P_B u$. Consequently if $v \in \mathbf{S}^s$ and $v \geq u$ on B

$$L(m, v) = L(m, \hat{v}) \geq L(m, P_B u) = \Gamma(B).$$

On the other hand let $D_B = \inf\{t \geq 0: X_t \in B\}$ be the entry time of B and let $H_B f = P_{D_B} f$. It is well-known and easily checked that $H_B u \in \mathbf{S}^s$, $(H_B u)\hat{} = P_B u$, and $H_B u = u$ on B. Therefore $\Gamma(B) = L(m, P_B u) = L(m, H_B u)$. ∎

(5.3) **Remarks.** Note that the proof shows that the infimum in (5.2) is actually attained when $v = H_B u$. Proposition 5.2 is false if $\mathbf{S}^s$ is replaced by $\mathbf{S}$. Let X be one dimensional Brownian motion killed when it exits $]-1, 1[$. Then $v(x) = 0$ for $x \neq 0$ and $v(0) = 1$ is supermedian but not strongly supermedian. Let m be Lebesgue measure on $]-1, 1[$, $u = 1$, and $B = \{0\}$. Then $v \geq u$ on B, and since $\hat{v} = 0$, $L(m, v) = 0$. But $\Gamma(B) = L(m, P_B 1) > 0$. Proposition 5.2 is also false if one replaces $\mathbf{S}^s$ by $\mathbf{E}$. Consider the example (3.10-i). For $B = \{0\}$, $\Gamma(B) = \frac{1}{2}$. But if $v \in \mathbf{E}$ and $v \geq 1$ on B, then (recall $m = \epsilon_0 U + \frac{1}{2}\lambda$)

$$L(m, v) \geq L(\epsilon_0 U, v) = v(0) = 1.$$

We turn next to a result which expresses Γ as an infimum over measures. For this we need the following result of Fitzsimmons [8].

(5.4) Theorem. *Let $m \in$ **Exc**. Then*

$$\inf\{\xi \in \mathrm{Exc}: \xi \geq m \quad \text{on} \quad B\} = R_{\sigma_B} m$$

where $\sigma_B = \inf\{t > \alpha: \int_\alpha^t 1_B \circ Y_s ds > 0\}$ is the Lebesgue penetration time of B for Y.

The following is an easy consequence of (5.4).

(5.5) Proposition. *Define*

$$\gamma(B) = \inf\{L(\xi, u) : \xi \in \mathrm{Exc},\ \xi \geq m \quad on \quad B\}.$$

Then $\Gamma(B) \geq \gamma(B) \geq L(R_{\sigma_B} m, u)$, *and so if* $R_{\sigma_B} m = R_B m$, *in particular if* B *is finely open, one has* $\gamma(B) = \Gamma(B)$.

Proof. Since $R_B m = m$ on B, $\gamma(B) \leq \Gamma(B)$. On the other hand (5.4) implies that $\gamma(B) \geq L(R_{\sigma_B} m, u)$. ∎

(5.6)**Remark.** Arguing exactly as in the proof of (3.16) of [14] one obtains

$$L(R_{\sigma_B} m, u) = L(m, P_{S_B} u).$$

The next result gives conditions under which one may approximate $\Gamma(B)$ by $\Gamma(G_n)$ wherc (G_n) is a decreasing sequence of finely open Borel supersets of B. The precise statement is Theorem 5.10. We shall need to use q-capacities for $q > 0$ and, in particular, the following result. However, we postpone its proof until section 6.

(5.7) Proposition. *Let* $m = m_i + m_p$ *be the decomposition of* m *into its invariant and purely excessive parts and let* $m_p = \int_0^\infty \eta_t dt$ *where* $\eta = (\eta_t)$ *is an entrance law for* (P_t). *Then for* $q > 0$

$$\Gamma^q(B) = qm(P_B^q u) + \uparrow \lim_{t \downarrow 0} e^{-qt} \eta_t (P_B^q u).$$

We introduce the following hypothesis on u:

(5.8) *Let* (T_n) *be an increasing sequence of stopping times and* $T = \lim T_n$ a.s. P^x. *Then for* $q > 0$, $\lim P_{T_n}^q u(x) = P_T^q u(x)$.

Remarks. Of course $P_{T_n}^q u$ decreases as n increases since $u \in \mathbf{E} \subset \mathbf{E}^q$. The function $u = 1$ satisfies (5.8) because $q > 0$. In general $T_n \uparrow T$ does not imply

$(q = 0)$, $P_{T_n}1 \to P_t 1$. If $u(x) = P^x(A_\infty) < \infty$ where $A = (A_t)$ is a continuous additive functional, then u satisfies (5.8) even for $q = 0$.

Finally we make the following assumption on m:

$$m = \nu U + \rho \qquad \text{with} \qquad \rho \in \text{Inv}. \tag{5.9}$$

This amounts to assuming that the harmonic part of m is invariant, or, equivalently, that the purely excessive part of m is a potential.

(5.10) Theorem. *Let u satisfy (5.8) and m be of the form (5.9). Suppose for some $p > 0$ there exists a finely open $G \in \mathcal{E}$ with $B \subset G$ and $\Gamma^p(G) < \infty$. If $\nu(B - B^r) = 0$, then there exists a decreasing sequence (G_n) of finely open Borel supersets of B such that $\Gamma^q(G_n) \downarrow \Gamma^q(B)$ for all $q \geq 0$.*

(5.11) Remark. If for some $p > 0$, $a > 0$, and $f \in p\mathcal{E}$ with $m(f) < \infty$, one has $B \subset \{U_p f > a\}$, then $G = \{U_p f > a\}$ satisfies the hypotheses in (5.10) as in the proof of (7.12) in [14].

Proof. Since $\nu(B - B^r) = 0$ by assumption and $m(B - B^r) = 0$ because $m \in \text{Exc}$ and $B - B^r$ is semipolar, there exists a decreasing sequence (G_n) of finely open Borel supersets of B with $T_{G_n} \uparrow T_B$ a.s. P^m and P^ν. See, for example, (12.10) of [10]. One may suppose $G_1 \subset G$ and so $\Gamma^p(G_1) < \infty$. Hence, by (8.1) of [14], $\Gamma^q(G_1) < \infty$ for all $q \geq 0$. Because of (5.9) one has $\eta_t = \nu P_t$ in (5.7) and so by (5.7) for each $q > 0$,

$$\Gamma^q(G_n) = qm(P^q_{G_n} u) + \nu(P^q_{G_n} u).$$

But $T_{G_n} \uparrow T_B$ a.s. P^x for $m + \nu$. a.e. x and hence by (5.8), $P^q_{G_n} u$ decreases to $P^q_B u$ as $n \to \infty$ a.e. $m + \nu$. Using $\Gamma^q(G_1) < \infty$, the dominated convergence theorem and (5.7) imply that $\Gamma^q(G_n) \downarrow \Gamma^q(B)$ for each $q > 0$. According to (8.1) of [14], $\Gamma^q(G_n) \downarrow \Gamma(G_n)$ and $\Gamma^q(B) \downarrow \Gamma(B)$ as $q \downarrow 0$. Therefore one may interchange the order of the limits to obtain $\Gamma(G_n) \downarrow \Gamma(B)$. ∎

(5.12) **Remarks.** The assumption (5.9) that the harmonic part of m is invariant may be eliminated by using the compactification methods in [11] and [12]. Moreover if one does *not* assume that $\nu(B - B^r) = 0$, then the same proof shows that for each $q \geq 0$

$$\Gamma^q(G_n) \downarrow \Gamma^q_{\rho,u}(B) + \nu(H^q_B u) \tag{5.13}$$

where, of course, $H^q_B = P^q_{D_B}$. The right hand side of (5.13) equals $\Gamma^q(B)$ if and only if $u = P^q_B u$ a.e. ν on $B - B^r$.

The final result in this section is an extension of (5.5) under stronger hypotheses. It should be compared to the results in [19]. We begin by defining for $B \in \mathcal{E}$

$$\gamma_1(B) = \inf \{L(\xi, u) : \xi \geq m \quad \text{on a fine neighborhood of} \quad B\},$$

$$\gamma_2(B) = \inf \{\mu(u) : \mu U \geq m \quad \text{on a fine neighborhood of} \quad B\},$$

where in both definitions a fine neighborhood of B means a finely open Borel set G containing B. Recall the definition of $\gamma(B)$ in (5.5).

(5.15) Proposition. *(i)* $\gamma(B) \leq \Gamma(B) \leq \gamma_1(B) \leq \gamma_2(B)$, *and* $\gamma(B) = \gamma_1(B)$ *if* B *is finely open.*

(ii) If there exists a decreasing sequence (G_n) *of fine neighborhoods of* B *with* $\Gamma(G_n) \downarrow \Gamma(B)$, *then* $\Gamma(B) = \gamma_1(B)$. *If, in addition, there exists a* $\mu U \in \text{Exc}$ *with* $\mu U \geq m$ *on a fine neighborhood of* B, *then* $\Gamma(B) = \gamma_2(B)$.

(5.16) **Remarks.** Of course, (5.10) gives conditions under which the first hypothesis in (5.15-ii) holds. It is the conjunction of (5.10) and (5.15) that gives results analogous to those in [19].

Proof. By (5.5), $\gamma(B) \leq \Gamma(B)$. Suppose G is a fine neighborhood of B and $\xi \in \text{Exc}$ with $\xi \geq m$ on G. Then $\Gamma(B) \leq \Gamma(G)$ and by (5.5) $\Gamma(G) \leq L(\xi, u)$. Consequently $\Gamma(B) \leq \gamma_1(B)$. Since $\mu(u) = L(\mu U, u)$, it is clear that $\gamma_1(B) \leq \gamma_2(B)$. Obviously $\gamma(B) = \gamma_1(B)$ if B is finely open. This establishes (i). If (G_n)

is a decreasing sequence of fine neighborhoods of B with $\Gamma(G_n) \downarrow \Gamma(B)$, then because $\Gamma(G_n) = \gamma(G_n) = \gamma_1(G_n)$ it is immediate that $\Gamma(B) = \gamma_1(B)$. Suppose $\mu_0 U \in \text{Exc}$ with $\mu_0 U \geq m$ on a fine neighborhood G of B. Choose $\xi_n \in \text{Exc}$ with $\xi_n \geq m$ on $G_n \supset B$ and $L(\xi_n, u) \downarrow \Gamma(B)$. Then $\eta_n = \xi_n \wedge (\mu_0 U) \in \text{Exc}$ and $\eta_n = \mu_n U$ since $\eta_n \leq \mu_0 U$. But $\eta_n \geq m$ on $G_n \cap G$ and $\mu_n(u) = L(\eta_n, u) \leq L(\xi_n, u)$. Therefore $\Gamma(B) = \gamma_2(B)$. ∎

6. Some Additional Formulas for Γ

We begin this section by proving Proposition 5.7. Fix $q > 0$ and let $\eta_t^q = e^{-qt}\eta_t + qmP_t^q$ where $P_t^q = e^{-qt}P_t$. Then $\eta^q = (\eta_t^q)_{t<0}$ is an entrance law for P_t^q. We claim that

$$m = \int_0^\infty \eta_t^q \, dt. \tag{6.1}$$

To this end, first note that $m_i = q\int_0^\infty m_i P_t^q dt$, and that

$$\begin{aligned} q\int_0^\infty m_p P_t^q dt &= q\int_0^\infty e^{-qt}\int_t^\infty \eta_s ds\ dt \\ &= \int_0^\infty \eta_s(1 - e^{-qs})ds = m_p - \int_0^\infty e^{-qs}\eta_s ds\ dt. \end{aligned}$$

Adding these relationships yields (6.1). These manipulations are easily justified by calculating $m(f)$ for $f \geq 0$ with $m(f) < \infty$. Using (6.1) we obtain

$$\begin{aligned} \Gamma^q(B) = L^q(m, P_B^q u) &= \uparrow \lim_{t\to 0} \eta_t^q(P_B^q u) \\ &= qm(P_B^q u) + \uparrow \lim_{t\to 0} e^{-qt}\eta_t(P_B^q u), \end{aligned}$$

since $mP_t^q \uparrow m$ as $t \downarrow 0$, and this is the statement of (5.7). Of course, one may drop the factor e^{-qt} in the term involving the limit, but then one does not have an increasing limit in general.

In order to state our next result, recall the definition of the birthing operators b_t defined in (5.9) of [14].

(6.2) Proposition. *If* $q > 0$, $\Gamma^q(B) = qQ_m^u(e^{-q\tau_B \circ b_0})$.

Proof. It actually suffices to prove this in the special case $u = 1$, but it is, perhaps, instructive to carry out the calculation for general u, and so we shall.

Since (Y_t) under Q_m^u is strong Markov with transition semigroup $P_t^{(u)}$—the u transform of P_t—we need to introduce some notation for u-transforms. See section 5 of [12], for example. Let $P^{x/u}$ be the basic probabilities on Ω under which (X_t) is Markov with semigroup $P_t^{(u)}$. Since m is fixed we may suppose that u is Borel measurable. Let T be an $(\mathcal{F}_{t+}^*)$ stopping time and $F \in p\mathcal{F}_{T+}^*$. Then

$$\text{(6.3)} \qquad u(x)P^{x/u}[F;\ T < \zeta] = P^x[Fu \circ X_T] \quad \text{if} \quad u(x) < \infty.$$

Since $m(u = \infty) = 0$, (6.3) holds a.e. m. Moreover if $m = m_i + \int_0^\infty \eta_t dt$ is the decomposition of m into its invariant and purely excessive parts as in (5.7), then using the fact that $1_{\{u=\infty\}}$ is supermedian (for (P_t)) it follows readily that (6.3) holds a.e. η_t for each $t > 0$. We are now ready to prove (6.2). Break $Q_m^u(e^{-q\tau_B \circ b_0})$ into integrals J_1 over $\{\alpha < 0\}$ and J_2 over $\{\alpha \geq 0\}$. Since $\tau_B \circ b_0 = T_B \circ \gamma_0$ on $\{\alpha < 0\}$,

$$\begin{aligned} J_1 &= Q_m^u(e^{-qT_B \circ \gamma_0} ; \alpha < 0) = Q_m^u[P^{Y(0)/u}(e^{-qT_B})] \\ &= \int m(dx)u(x)P^{x/u}(e^{-qT_B}) = mP_B^q u, \end{aligned}$$

where the last equality follows from (6.3) and the fact that $T_B < \zeta$ if $T_B < \infty$.

Next $\tau_B \circ b_0 = \tau_B$ on $\{0 < \alpha\}$ and $Q_m^u(\alpha = 0) = 0$, so $J_2 = Q_m^u(e^{-q\tau_B};\ 0 < \alpha)$. Now Q_m^u is the Kuznetsov measure corresponding to um, 1, and $P_t^{(u)}$ and according to (4.4) of [14], $(um)_p = um_p$. But $(u\eta_t)$ is an entrace law for $P_t^{(u)}$ and $um_p = \int_0^\infty u\eta_t\, dt$. If we write $\tilde{Q}$ for Kuznetsov measures corresponding to the semigroup $(P_t^{(u)})_{t>0}$, then

$$\begin{aligned} J_2 &= \tilde{Q}_{um}(e^{-q\tau_B};\ 0 < \alpha) = \tilde{Q}_{(um)_p}(e^{-q\tau_B};\ 0 < \alpha) \\ &= \int_{-\infty}^{\infty} \tilde{Q}_{u\eta}(e^{-q\tau_B \circ \theta_t};\ 0 < \alpha \circ \theta_t)dt \\ &= \int_{-\infty}^{\infty} \tilde{Q}_{u\eta}(e^{-q(\tau_B - t)};\ t < \alpha)dt. \end{aligned}$$

But $\alpha = 0$ a.s. $\tilde{Q}_{u\eta}$ since $u\eta$ is an entrance law for $P_t^{(u)}$, and so

$$J_2 = \tilde{Q}_{u\eta}\left(e^{-q\tau_B}\int_{-\infty}^{0} e^{qt}\,dt\right) = q^{-1}\tilde{Q}_{u\eta}(e^{-q\tau_B}).$$

Observe that $t + T_B \circ \gamma_t \downarrow \tau_B$ a.s. $\tilde{Q}_{u\eta}$ as $t \downarrow 0$. Hence

$$\begin{aligned} qJ_2 &= \lim_{t\downarrow 0} e^{-qt}\tilde{Q}_{u\eta}(e^{-qT_B\circ\gamma_t}) \\ &= \lim_{t\downarrow 0} e^{-qt}\tilde{Q}_{u\eta}[P^{Y(t)/u}(e^{-qT_B})] \\ &= \lim_{t\downarrow 0} e^{-qt}\int \eta_t(dx)u(x)P^{x/u}(e^{-qT_B}) \\ &= \lim_{t\downarrow 0} e^{-qt}\eta_t(P_B^q u), \end{aligned}$$

where the last equality follows because (6.3) holds a.e. η_t for each $t > 0$. Combining these calculations with (5.7) yields (6.2). ∎

Remark. One may also prove (6.2) first and then use it to derive (5.7).

Our last result was essentially proved in [14], but seems interesting enough to make explicit. For simplicity we take $u = 1$, but the analogous formula holds for general u. Formula (8.9) of [14] may then be written

$$\Gamma^q(B) = \Gamma(B) + qM(B) + \int_0^\infty (1 - e^{-qt})\nu_B(dt) \tag{6.4}$$

where

$$\nu_B(dt) = {}^*P^\nu(T_B \in dt,\ T_B < \infty) \tag{6.5}$$

is the distribution of T_B under the excursion law ${}^*P^\nu$ and

$$M(B) = Q_m(G_0 = 0 = T_B \circ \gamma_0,\ \tau_B < 0 < \lambda_B) \tag{6.6}$$

and G_0 is defined below (8.5) in [14]. Let $B^f = B \cup B^r$ be the fine closure of B. Then $\Gamma^q(B) = \Gamma^q(B^f)$ for all $q \geq 0$. For simplicity we shall suppose that B is finely closed. Then $G_0 = \sup\{t \leq 0 \colon Y_t \in B\}$ and on $\{\tau_B < 0 < \lambda_B\} \subset \{\alpha < 0 < \beta\}$ one has a.s. Q_m

$$\{T_B \circ \gamma_0 = 0\} = \{Y_0 \in B^r\} \subset \{Y_0 \in B\} \subset \{G_0 = 0\}.$$

Also if $Y_0 \in B^r$, then $\tau_B \leq 0 \leq \lambda_B$ and so

$$M(B) = Q_m(Y_0 \in B^r) = m(B^r) = m(B).$$

Thus we have the following result.

(6.7) Theorem. *Let $u = 1$ and B be finely closed, then*

$$\Gamma^q(B) = \Gamma(B) + qm(B) + \int_0^\infty (1 - e^{-qt})\nu_B(dt) \tag{6.8}$$

where ν_B is given by (6.5).

It follows from (6.8) that if $\Gamma^q(B)$ is finite for one $q > 0$, then $q \to \Gamma^q(B)$ is a subordinator exponent, that is, it has a completely monotone derivative. Moreover under this condition $\Gamma^q(B) \downarrow \Gamma(B)$ as $q \downarrow 0$ and $q^{-1}\Gamma^q(B) \downarrow m(B)$ as $q \to \infty$. The first example in (8.3) of [14] shows that one may have $\Gamma(B) = 0$ and $\Gamma^q(B) = \infty$ for all $q > 0$. Also if X is translation to the right on $\mathbf{R}$ at unit speed with m Lebesgue measure and $B = \mathbf{Z}$ (the integers), then $\Gamma^q(B) = \infty$ for $q > 0$ and $m(B) = 0$. Thus the assumption $\Gamma^q(B) < \infty$ for some $q > 0$ is necessary for the limiting relations

$$\downarrow \lim_{q \to 0} \Gamma^q(B) = \Gamma(B); \downarrow \lim_{q \to \infty} q^{-1}\Gamma^q(B) = m(B). \tag{6.9}$$

References

1. R. M. Blumenthal and R. K. Getoor. *Markov Processes and Potential Theory.* Academic Press, New York. 1968.
2. R. M. Blumenthal and R. K. Getoor. Dual processes and potential theory. Proc. 12th Biennial Sem. Can. Math. Soc., 137–156. 1970.
3. Z. Ciesielski. Semiclassical potential theory. Markov Processes and Potential Theory, edited by J. Chover. 33–60. Wiley and Sons. New York, London, Sydney. 1967.
4. C. Dellacherie and P. A. Meyer. Probabilités et Potentiel. Ch. XII á XVI. Hermann. Paris. 1987.

5. P. J. Fitzsimmons. On two results in the potential theory of excessive measures. Sem. Stoch. Proc. 1986, 21–30. Birkhäuser. Boston. 1987.

6. P. J. Fitzsimmons. Homogeneous random measures and a weak order for the excessive measures of a Markov process. To appear in Trans. Amer. Math. Soc.

7. P. J. Fitzsimmons. On a connection between Kuznetsov processes and quasi-processes. To appear in Sem. Stoch. Proc. 1987. Birkhäuser, Boston.

8. P. J. Fitzsimmons. Penetration times and Skorohod stopping. To appear in Sém. de Probabilités XXII. Lec. Notes in Math. Springer. Berlin-Heidelberg-New York.

9. P. J. Fitzsimmons and B. Maisonneuve. Excessive measures and Markov processes with random birth and death. Probab. Th. Rel. Fields **72**, 391–336 (1986).

10. R. K. Getoor. Markov Processes: Ray Processes and Right Processes. Lecture Notes in Math. **40**, Springer. Berlin-Heidelberg-New York. 1975.

11. R. K. Getoor and J. Glover. Markov processes with identical excessive measures. Math. Zeit. **184**, 287–300 (1983).

12. R. K. Getoor an J. Glover. Riesz decompositions in Markov process theory. Trans. Amer. Math. Soc. **285** 107–132 (1984).

13. R. K., Getoor and J. Steffens. Capacity theory without duality. Probab. Th. Rel. Fields **73** 415–445 (1986).

14. R. K. Getoor and J. Steffens. The energy functional, balayage, and capacity. Ann. Inst. Henri Poincaré **23** 321–357 (1987).

15. J. Glover. Topics in energy and potential theory. Sém. Stoch. Proc. 1982, 195–202. Birkhäuser, Boston, 1983.

16. G. A. Hunt. Markov processes and potentials I. Ill. J. Math. **1** 44–93 (1957).

17. D. Revuz. Remarque sur les potentiels de mesure. Sém. de Prob V. Lecture Notes in Math. **191** 275–277. Springer. Berlin-Heidelberg-New York. 1971.

18. D. Stroock. The Kac approach to potential theory I and II. J. Math. Mech. **16**, 829–852 (1967) and Comm. Pure Appl. Math. **20**, 775–796 (1967).

19. B. Fuglede. Le théorème du minimax et la théorie fine du potentiel. Ann. Inst. Fourier, Grenoble. **15**, 65–88 (1965).

R. K. Getoor

Department of Mathematics, C-012

University of California, San Diego

La Jolla, California 92093

J. Steffens

Institut für Statistik und Documentation

Universität Düsseldorf

D-4000 Düsseldorf 1

West Germany

CAPACITIES OF SYMMETRIC MARKOV PROCESSES

by

J. GLOVER*, W. HANSEN, and M. RAO

0. INTRODUCTION

Let $X = (\Omega, \mathcal{F}, \mathcal{F}_t, X_t, \theta_t, P^x)$ and $Y = (\Omega, \mathcal{G}, \mathcal{G}_t, Y_t, \theta_t, Q^x)$ be two transient Hunt processes on an LCCB state space E with Borel field $\mathcal{E}$. Let $T_K = \inf\{t>0: X_t \in K\}$, and let $S_K = \inf\{t>0: Y_t \in K\}$. The following result ([3], [4], [5], [6]) is a strengthening of the well-known theorem of Blumenthal, Getoor, and McKean ([1], V-5.1).

(0.1) THEOREM. Assume $P^x(T_K<\infty) = Q^x(S_K<\infty)$ for every $K \in \mathcal{E}$ and for every $x \in E$. There is a strictly increasing continuous additive functional A_t of X_t so that if $\tau_t = \inf\{s: A_s>t\}$, then $(X(\tau_t), P^x)$ and (Y_t, Q^x) are identical in law.

In the language of potential theory, (0.1) states that the regularized reduites of the function 1 on all Borel sets determines the potential theory. Blumenthal, Getoor, and McKean showed that two processes with the same hitting

*Research supported in part by AFOSR Grant 85-0330 and NSF Grant DMS-8318204.

distributions are time changes of one another, and Theorem (0.1) states that two processes with the same hitting probabilities are time changes of one another. Can we obtain the same conclusion with even less information?

To continue, we assume a mild regularity hypothesis. Assume that we have two more Hunt processes $\hat{X}$ and $\hat{Y}$ on E and measures m and n so that X and $\hat{X}$ are in duality with respect to m and Y and $\hat{Y}$ are in duality with respect to n as described in Chapter VI of [1]. (This hypothesis is automatically satisfied if E is finite.) Let $u(x,y)$ and $v(x,y)$ be the potential densities of X and Y. If K is a transient set for X and Y, then $P^x(T_K<\infty) = U\pi_K(x)$ and $Q^x(S_K<\infty) = V\gamma_K(x)$, where π_K and γ_K are the appropriate equilibrium measures of the set K, and where

$$U\pi_K(x) = \int_E u(x,y)\pi_K(dy),$$

$$V\gamma_K(x) = \int_E v(x,y)\gamma_K(dy).$$

Instead of assuming that $U\pi_K = V\gamma_K$ as in Theorem (0.1), we assume only that $\pi_K(E) = \gamma_K(E)$. That is, we are assuming that the capacities generated by the two processes agree. (Recall that $\pi_K(E)$ is the capacity of K.) If we take two sequences of measures (λ_k) and (μ_k) so that $\int u(y,x)\lambda_k(dy)$ and $\int v(y,x)\mu_k(dy)$ increase to the constant function 1 on E, then

$$\pi_K(E) = \lim_{k\to\infty} P^{\lambda_k}(T_K<\infty)$$

$$\gamma_K(E) = \lim_{k\to\infty} Q^{\mu_k}(S_K<\infty).$$

Therefore, by assuming that $\pi_K(E) = \gamma_K(E)$, we are assuming less than in (0.1). Can we conclude that X and Y must be time changes of one another? Alas, the answer is clearly no. Theorem (VI-4.4) in [1] states that a process and its dual have the same capacities, and, in general, a process and its dual are not time changes of one anther. But if we insist that X and Y be symmetric ($u(x,y) = u(y,x)$ and $v(x,y) = v(y,x)$), then we do obtain that X and Y are time changes of one another (provided we assume a mild continuity hypothesis.) In fact, we give two remarkable formulas for $u(x,y)$ in terms of the capacity. If the process hits points, the formula (0.3) involves only the capacities of one- and two-point sets. If points are polar for the process, the formula (0.5) gives $u(x,y)$ as a limit of capacities.

(0.2) THEOREM. Assume that $u(z,w) = u(w,z)$ and $P^z(T_{\{z\}}<\infty)>0$ for all z and w in E. Fix x and y in E, let a be the capacity of $\{x\}$, let b be the capacity of $\{y\}$, and let c be the capacity of the set $\{x,y\}$. Then

$$(0.3)\qquad u(x,y) = \frac{1 - \sqrt{1 - c[1/a + 1/b - c/ab]}}{c}$$

REMARK. Because of the symmetry, the assumption $P^z(T_{\{z\}}<\infty) > 0$ implies, in fact, that $P^z(T_{\{z\}} =0) = 1$: every point in E is regular for X. In particular, $u(z,z) < \infty$.

In order to state the result in the case that points are polar, we need to introduce some notation. Let x and y be two points in E, and let G and H be relatively compact open neighborhoods of x and y. Define

$$F_{G,H} = \{(A,B) \in \mathscr{E} \times \mathscr{E} : A\times B \subset G\times H,\ \pi_A(E) = \pi_B(E) > 0\}.$$

By Proposition (2.2), $F_{G,H} \neq \emptyset$. If we let $\mathscr{H}(x)$ be a fundamental system of relatively compact neighborhoods of x, then $\{F_{G,H}: G \in \mathscr{H}(x),\ H \in \mathscr{H}(y)\}$ is the base of a filter $\mathscr{F}_{x,y}$ on $\{(A,B) \in \mathscr{E} \times \mathscr{E} : \pi_A(E) = \pi_B(E) > 0\}$.

(0.4) THEOREM. Assume that $u(z,w) = u(w,z)$ and $P^z(T_{\{w\}} < \infty) = 0$ for all z and w in E. Let $(x,y) \in E \times E$ such that $u(x,y) < \infty$ and u is continuous at (x,y) in the product topology on $E \times E$. Let $c(A) = \pi_A(E)$ whenever $A \in \mathscr{E}$ is relatively compact. Then

$$(0.5) \qquad u(x,y) = \lim_{(A,B),\mathscr{F}_{x,y}} \left(\frac{2}{c(A \cup B)} - \frac{1}{c(A)}\right).$$

The proof of Theorem (0.2) is given in section 1, and several consequences and corollaries are discussed. The proof of Theorem (0.4) is given in section 2. After we wrote this article, A. Ancona pointed out to us that Theorem (0.4) is contained in Choquet [2]. We are continuing to print this for two reasons: (1) Theorem (0.4) is not well-known to probabilists, and (2) the discrete case (0.2) is not contained in Choquet and the formula and proof are a bit different.

1. POINTS ARE REGULAR.

Before giving the proof of Theorem (0.2), we observe several consequences.

(1.1) COROLLARY. *Let* X *and* Y *be transient symmetric processes which hit points and which have the same capacities. Then* X *and* Y *are time changes of one another.*

PROOF. By (0.3) X and Y have the same potential density, so they have the same cone of excessive functions, and hence the same hitting distributions. The result now follows from the Blumenthal, Getoor, McKean theorem. Alternatively, just let A_t be the continuous additive functional of X_t with Revuz measure n. If $\tau_t = \inf\{s: A_s > t\}$, then it is well-known that $X(\tau_t)$ is a symmetric process with potential density u(x,y) and duality measure n, and hence must be identical in law to Y_t. Q.E.D.

The next result is immediately obvious from formula (0.3).

(1.2) COROLLARY. *The capacities of one- and two-point sets determine the potential theory of symmetric processes which hit points.*

REMARK. Formula (0.3) simplifies nicely if the capacities of one- or two-point sets are translation invariant on a locally compact abelian group E. Suppose $a = \pi_{\{x\}}(E)$ is independent of x. Then

$$u(x,y) = \frac{1 - |c/a - 1|}{c} = \frac{2}{c} - \frac{1}{a},$$

since $c/a > 1$. This is reminiscent of formula (0.5). If $c = c(x,y)$ depends only on the distance between x and y, then we see that u(x,y) is translation invariant on E, so X is a time change of a Levy process.

We use the notation given in (0.2) in the proof.

PROOF of (0.2). If x=y, then a=b=c, and (0.3) becomes $u(x,x) = a^{-1}$, which is true since $u(x,x)a = 1$. Now assume $x \neq y$, and let π be the equilibrium measure of the two-point set $\{x,y\}$. Since points are regular, we have $U\pi(x) = U\pi(y) = 1$. If we use the notations $\pi_x = \pi(\{x\})$, $\pi_y = \pi(\{y\})$, $u_{xx} = u(x,x)$, $u_{xy} = u_{yx} = u(x,y)$ and $u_{yy} = u(y,y)$, then these two equations become

$$(1.3)\qquad \begin{aligned} u_{xx}\pi_x + u_{xy}\pi_y &= 1 \\ u_{yx}\pi_x + u_{yy}\pi_y &= 1. \end{aligned}$$

The maximum principles states that $u_{xy} \leq u_{yy}$ and $u_{yx} \leq u_{xx}$. Adding the equations (1.3) and using $u_{yx} = u_{xy}$ we obtain that $2u_{xy}(\pi_x+\pi_y) \leq 2$, i.e.

$$(1.4)\qquad u_{xy}c \leq 1.$$

Suppose for the moment that $u_{xx}u_{yy} \neq u_{xy}^2$. Then (1.3) implies that

$$\pi_x = \frac{u_{yy} - u_{xy}}{u_{xx}u_{yy} - u_{xy}^2}$$

and

$$\pi_y = \frac{u_{xx} - u_{xy}}{u_{xx}u_{yy} - u_{xy}^2},$$

hence,

$$c = \frac{u_{xx} + u_{yy} - 2u_{xy}}{u_{xx}u_{yy} - u_{xy}^2}.$$

By clearing the denominator and rearranging terms, we obtain the following equation:

$$cu_{xy}^2 - 2u_{xy} + u_{xx} + u_{yy} - cu_{xx}u_{yy} = 0.$$

or

$$cu_{xy}^2 - 2u_{xy} + a^{-1} + b^{-1} - ca^{-1}b^{-1} = 0.$$

In consideration of (1.4) we thus conclude that

$$u_{xy} = \frac{1 - \sqrt{1 - c(a^{-1} + b^{-1} - ca^{-1}b^{-1})}}{c}.$$

So the proof is finished if $u_{xx}u_{yy} \neq u_{xy}^2$. Suppose finally that $u_{xx}u_{yy} = u_{xy}^2$. Then either $u_{xy} > u_{xx}$ or $u_{xy} > u_{yy}$. Say $u_{xy} > u_{yy}$. By the maximum principle $u_{xy} = u_{yy}$. Therefore $u_{xy}c = u_{yy}c = 1$ by the second equation in (1.3). Hence $c = b$ and

$$u_{xy} = \frac{1}{c} = \frac{1 - \sqrt{1 - c(a^{-1}+b^{-1}-ca^{-1}b^{-1})}}{c}.$$

The argument for the case $u_{xy} > u_{xx}$ is exactly the same. Q.E.D.

2. POINTS ARE POLAR.

In this section, we use the notation and hypotheses of (0.4), so $c(A)$ is the capacity of the set A. Also, since points are polar, $c(\{x\}) = 0$ for every x in E. Let $\mathcal{M}^+$ denote the collection of positive σ-finite measures on

$(E, \mathscr{E})$. Our proof of (0.4) will be based on the following characterization of $c(A)$: For every $A \in \mathscr{E}$

(2.1) $$c(A) = \sup\{\mu(E): \mu \in \mathscr{M}^+, U\mu \leq 1, \mu(A^c) = 0\}.$$

Let us establish first that $F_{G,H} \neq \emptyset$.

(2.2) PROPOSITION. Let $A \in \mathscr{E}$, and let $0 < \alpha \leq c(A)$. There is a Borel subset B of A so that $c(B) = \alpha$.

PROOF. If $\alpha = c(A)$, take $B=A$. If $\alpha < c(A)$, there is a relatively compact subset A' of A so that $\alpha < c(A')$. Therefore we may assume that A is relatively compact by replacing A with A'. We now prove that there is an increasing sequence $(B_n) \subset \mathscr{E}$ so that for every n,

(2.3) $$B_n \subset A \text{ and } (1 - 1/n)\alpha < c(B_n) \leq \alpha.$$

Let $B_1 = \emptyset$, and assume we have chosen B_n so that (2.3) holds. Since points are polar and A is relatively compact, we may choose Borel subsets $A_1, A_2, \ldots, A_k$ of $A-B_n$ so that

$$A - B_n = \bigcup_{i=1}^{k} A_i,$$

and $c(A_i) < \alpha/(n+1)$ for every $i \leq k$. Since $c(B_n) \leq \alpha$ and $c(B_n \cup (A-B_n)) = c(A) > \alpha$, there is a j with $1 \leq j \leq k$ so that

$$c(B_n \cup \bigcup_{i=1}^{j-1} A_i) \leq \alpha \leq c(B_n \cup \bigcup_{i=1}^{j} A_i).$$

Define

$$B_{n+1} = B_n \cup \bigcup_{i=1}^{j-1} A_i .$$

Then

$$c(B_{n+1}) \leq \alpha < c(B_{n+1} \cup A_j) \leq c(B_{n+1}) + c(A_j)$$

$$< c(B_{n+1}) + \alpha/(n+1).$$

Thus

$$\alpha - \frac{\alpha}{n+1} < c(B_{n+1}) \leq \alpha .$$

Finally, let

$$B = \bigcup_{n=1}^{\infty} B_n .$$

then $B \in \mathscr{E}$, $B \subset A$, and $c(B) = \sup c(B_n) = \alpha$. Q.E.D.

In view of (2.2), Theorem (0.4) is an immediate consequence of the following proposition.

(2.3) PROPOSITION. <u>Let</u> $(x,y) \in E \times E$ <u>with</u> $u(x,y) < \infty$. <u>Let</u> $\varepsilon > 0$, <u>and let</u> $A,B \in \mathscr{E}$ <u>satisfy</u> $0 < c(A) = c(B) < \infty$ <u>and</u> $|u(x',y') - u(x,y)| \leq \varepsilon$ <u>for all</u> $x' \in A$, $y' \in B$. <u>Then</u>

$$\left| u(x,y) - \frac{2}{c(A \cup B)} - \frac{1}{c(A)} \right| \leq \varepsilon .$$

PROOF. Let $\alpha = c(A) = c(B)$, and let $\gamma = c(A \cup B)$. Then $0 < \alpha \leq \gamma \leq 2\alpha < \infty$. Fix $\delta > 0$. By (2.1) there exists $\mu \in \mathscr{M}^+$ such that $U\mu \leq 1$, $\mu(A^c) = 0$, and $\mu(E) \leq \alpha < \mu(E) + \alpha\gamma\delta/2$. For every $y' \in B$,

$$U\mu(y') = \int u(y',x')\mu(dx') = \int u(x',y')\mu(dx')$$
$$\leq (u(x,y)+\varepsilon)\mu(E) \leq (u(x,y)+\varepsilon)\alpha .$$

In a similar manner, by choosing $\nu \in \mathscr{M}^+$ so that $U\nu \leq 1$, $\nu(B^c) = 0$, and $\nu(E) \leq \alpha < \nu(E) + \alpha\gamma\delta/2$, we obtain

$$U\upsilon(y') \leq (u(x,y)+\varepsilon)\alpha$$

for every $y' \in A$. Therefore,

$$U(\mu+\upsilon) \leq 1 + (u(x,y)+\varepsilon)\alpha \text{ on } A \cup B.$$

Defining the measure

$$\lambda = [1 + (u(x,y)+\varepsilon)\alpha]^{-1}(\mu+\upsilon),$$

we have $\lambda((A \cup B)^c) = 0$ and $U\lambda \leq 1$ (since $U\lambda \leq 1$ on $A \cup B$). Thus we conclude by (2.1) that

$$\gamma = c(A \cup B) \geq \lambda(E) \geq \frac{2\alpha - \alpha\gamma\delta}{1 + (u(x,y)+\varepsilon)\alpha}$$

or

$$u(x,y) \geq \frac{2}{\gamma} - \frac{1}{\alpha} - \varepsilon - \delta.$$

Next fix $0 < \delta' < \gamma$ such that $2/(\gamma-\delta') \leq (2/\gamma)+\delta$. There exists $\rho \in \mathcal{M}^+$ such that $U\rho \leq 1$, $\rho((A \cup B)^c) = 0$ and $\rho(E) \leq \gamma < \rho(E) + \delta'$. Let $\sigma = 1_A\rho$ and $\tau = 1_B\rho$. Then $\sigma(A^c) = \tau(B^c) = 0$ and $\sigma + \tau = \rho$. Clearly, $U\tau \geq (u(x,y)-\varepsilon)\tau(E)$ on A, and hence

$$U\sigma = U\rho - U\tau \leq 1 - (u(x,y)-\varepsilon)\tau(E) \text{ on } A.$$

Since $\sigma(A^c) = 0$, the preceding inequality holds on E. Hence by (2.1), $\alpha \geq [1 - (u(x,y)-\varepsilon)\tau(E)]^{-1}\sigma(E)$, or

$$\alpha[1 - (u(x,y)-\varepsilon)\tau(E)] > \sigma(E).$$

Similarly,

$$\alpha[1 - (u(x,y)-\varepsilon)\sigma(E)] > \tau(E).$$

By adding, we obtain

$$\alpha[2 - (u(x,y)-\varepsilon)\rho(E)] > \rho(E).$$

or

$$u(x,y) < \frac{2}{\rho(E)} - \frac{1}{\alpha} + \varepsilon < \frac{2}{\gamma-\delta'} - \frac{1}{\alpha} + \varepsilon < \frac{2}{\gamma} - \frac{1}{\alpha} + \varepsilon + \delta.$$

Q.E.D.

REMARK. Theorem (0.4) has been discussed in the context of a Markov process, and in this situation, a simpler proof of (2.3) can be obtained by using equilibrium measures. However, the proofs given above show that (2.2) and (2.3) hold if we only assume the following:

(i) $u\colon E\times E \to [0,\infty]$ is a symmetric Borel function which is infinite on the diagonal and such that $U\mu \le 1$ on E whenever $\mu \in \mathcal{M}^+$ and $A \in \mathcal{E}$ with $\mu(A^c) = 0$ and $U\mu \le 1$ on A.

(ii) c is a capacity on E asociated with u by (2.1).

So a formula of the type (0.5) is valid in a more general setting.

REFERENCES

1. R. M. Blumenthal and R. K. Getoor, Markov Processes and Potential Theory. Academic Press, New York (1968).

2. G. Choquet, Theory of capacities. Ann. Inst. Fourier 5(1955) 131-295.

3. P. Fitzsimmons, Markov Processes with identical hitting probabilities. (preprint).

4. J. Glover. Markov processes with identical hitting probabilities. Trans. AMS 275 (1983) 131-141.

5. J. Glover, Identifying Markov processes up to time change. Seminar on Stochastic Processes 1982, Boston, Birkhauser, (1983) 171-194.

6. D. Heath, Skorokhod stopping via potential theory. Seminaire de Probabilites VIII. Springer-Verlag, Berlin (1974) 150-154.

SOBOLEV SPACES, KAC-REGULARITY, AND THE FEYNMAN-KAC FORMULA

Ira W. Herbst* and Zhongxin Zhao

I. INTRODUCTION

Let $H^1(\mathbb{R}^d)$ be the Sobolev space of complex-valued measurable functions $u \in L^2(\mathbb{R}^d)$ such that the distributional derivatives $\partial_j u$, $j = 1,2,\cdots,d$ are also in $L^2(\mathbb{R}^d)$. We introduce the sesquilinear form $(\cdot,\cdot)_*$ on $H^1(\mathbb{R}^d)$:

$$(u,v)_* = \frac{1}{2}\int_{\mathbb{R}^d} \sum_j \partial_j \bar{u} \partial_j v \, dx$$

and the norm $\|\cdot\|$

$$\|u\|^2 = (u,v)_* + \|u\|_2^2$$

where $\|\cdot\|_2$ is the $L^2(\mathbb{R}^d)$-norm.

For any open set $D \subset \mathbb{R}^d$, let $C_0^\infty(D)$ denote the set of all infinitely differentiable functions on $\mathbb{R}^d$ with compact support in D. (Note the slight departure from the usual convention: If $f \in C_0^\infty(D)$, then f is defined everywhere in $\mathbb{R}^d$.) Corresponding to the open set D, we have the usual Sobolev space

$$H_0^1(D) = \text{closure of } C_0^\infty(D) \text{ in the norm } \|\cdot\|$$

For any Borel set M, let

$$L^2(M) = \{f \in L^2(\mathbb{R}^d) : f = 0 \ \ m\text{-a.e. in } M^c\}$$

where m is Lebesgue measure. We will consider the Sobolev space

$$\tilde{H}_0^1(M) = H^1(\mathbb{R}^d) \cap L^2(M).$$

Clearly, for any open set D,

$$H_0^1(D) \subset \tilde{H}_0^1(D) \subset \tilde{H}_0^1(\overline{D}) \tag{1.1}$$

One of the primary purposes of this paper is to characterize the open sets D for which each inclusion in (1.1) becomes equality.

These Sobolev spaces arise naturally from the consideration of Brownian motion and related semigroups. We will need the following general result. (See Kato [6] or Fukushima [5]).

THEOREM A: *There is a 1-1 correspondence between each of the following three families in a Hilbert space* $\mathcal{H}$:

(a) *non-negative, closed, sesquilinear forms* Q *which are densely defined;*

(b) *self-adjoint operators* A *which are non-positive:*

$$(Au, u) \leq 0, \qquad u \in \mathcal{D}(A);$$

(c) *strongly continuous semigroups of self-adjoint operators* $\{T_t\}_{t\geq 0}$ *with* $T_0 = I$.

The correspondence is determined by

$$Q(u,v) = (\sqrt{-A}\, u, \sqrt{-A}\, v)$$

with $\mathcal{D}(Q) = \mathcal{D}(\sqrt{-A})$ *and* A *is the infinitesimal generator of* $\{T_t\}$, $T_t = e^{At}$. ($\mathcal{D}(\cdot)$ *denotes the domain of an operator or form.*)

For any open set $D \subset \mathbb{R}^d$ we use the symbol Δ_D to denote the self-adjoint operator in $L^2(D)$ corresponding to the form

$$Q(u,v) = (u,v)_*, \quad \mathcal{D}(Q) = H_0^1(D).$$

Note that Δ_D is half the usual Dirichlet Laplacian. We use the notation 1_M for the function which is one on M and zero on M^c.

Our original motivation came from questions concerning the convergence of certain sequences of operators in $L^2(\mathbb{R}^d)$. Let $\{D_n\}_{n=1}^{\infty}$ be a sequence of open sets with $D_n \downarrow \overline{D}$ where D is open. Later we will prove the following theorem.

THEOREM 1.1: *For* $t > 0$, *the following sequences of operators converge strongly on* $L^2(\mathbb{R}^d)$ *as* $n \to \infty$:

$$\begin{aligned} \exp(t(\Delta - n1_{D^c})) &\longrightarrow P_1^t \\ (\exp(t\Delta/n)1_D)^n &\longrightarrow P_2^t \\ \exp(t(\Delta - n1_{\overline{D}^c})) &\longrightarrow P_3^t \\ (\exp(t\Delta/n)1_{\overline{D}})^n &\longrightarrow P_4^t \\ \exp(t\Delta_{D_n})1_{D_n} &\longrightarrow P_5^t \end{aligned} \tag{1.2}$$

Here $\{P_j^t\}_{t\geq 0}$ *is a strongly continuous semigroup of operators in* $L^2(\mathbb{R}^d)$ and $\Delta = \Delta_{\mathbb{R}^d}$. *The operator* P_j^0 *is an orthogonal projection, not necessarily the identity. We have*

$$P_1^t = P_2^t = e^{G_1 t} P_1^0$$

$$P_3^t = P_4^t = P_5^t = e^{G_2 t} P_3^0$$

where $\mathcal{D}(\sqrt{-G_1}) = \tilde{H}_0^1(D)$, $\mathcal{D}(\sqrt{-G_2}) = \tilde{H}_0^1(\overline{D})$, *and*

$$\|\sqrt{-G_j}\, u\|^2 = \sum_{k=1}^{d} \int |\partial_k u|^2 dx$$

for $u \in \mathcal{D}(\sqrt{-G_j})$.

In view of this result, questions about convergence in (1.2) lead directly to questions about the Sobolev spaces in (1.1) and vice versa.

In Section II, we use Brownian motion to represent the semigroups P_j^t. We recall the notion of Kac-regularity (first discussed by Stroock [11]) and give necessary and sufficient conditions for the equality of pairs of Sobolev spaces in (1.1) using ideas from Brownian motion (see Theorems 2.1 and 2.4). Independent of probabilistic ideas, a well-known restriction on D, the segment condition, is shown in Theorem 2.6 to be sufficient for the equality of all three Sobolev spaces.

Acknowledgement

We would like to thank Loren Pitt for bringing Stroock's paper [11] to our attention.

II. Sobolev Spaces, Quadratic Forms, and Semigroups

We will make use of a general theorem concerning the convergence of monotone increasing sequences of forms [4], [10].

THEOREM B: *Suppose* $\{Q_n\}_{n=1}^{\infty}$ *is a sequence of non-negative, closed, sesquilinear forms in a Hilbert space* $\mathcal{H}$. *For each* $n \geq 1$, *suppose*

$$Q_{n+1}(u,u) \geq Q_n(u,u), \quad u \in \mathcal{D}(Q_{n+1}) \subset \mathcal{D}(Q_n).$$

Then we can define a non-negative, closed form:

$$Q_\infty(u,v) = \lim_{n\to\infty} Q_n(u,v)$$

$$\mathcal{D}(Q_\infty) = \{u \in \bigcap_{n=1}^{\infty} \mathcal{D}(Q_n): \sup_n Q_n(u,u) < \infty\}.$$

Let A_n $(n = 1,2,\cdots,\infty)$ *denote the self-adjoint operator corresponding to* Q_n *as in Theorem A. Then* A_n *converges to* A_∞ *in the strong resolvent sense. Thus, for any continuous function* g *on* $(-\infty,0]$ *which vanishes at infinity, we have (strongly on* $\mathcal{H}$*)*

$$g(A_n)P_n \longrightarrow g(A_\infty)P_\infty \tag{2.1}$$

where P_n *is the orthogonal projection on the subspace*

$$\mathcal{H}_n = \overline{\mathcal{D}(Q_n)}, \quad n = 1,2,\cdots,\infty.$$

We summarize the convergence of the sequence of forms above by writing

$$-A_n \uparrow -A_\infty \tag{2.2}$$

Of particular interest to us is that, given (2.2), we have (as a special case of (2.1))

$$e^{tA_n} P_n f \to e^{tA_\infty} P_\infty f$$

for each $f \in \mathcal{H}$ and $t > 0$.

A related theorem concerning the approximation of semigroups will be of interest to us. It is a generalization of the Trotter product formula due to Kato and Masuda [7]:

THEOREM C: *Suppose Q_1 and Q_2 are non-negative, closed forms in a Hilbert space $\mathcal{H}$. Define a form Q_3:*

$$\mathcal{D}(Q_3) = \mathcal{D}(Q_1) \cap \mathcal{D}(Q_2)$$

$$Q_3(u,v) = Q_1(u,v)+Q_2(u,v); \quad u,v \in \mathcal{D}(Q_3)$$

Let A_j be the self-adjoint operator corresponding to the form Q_j in the Hilbert space $\mathcal{H}_j = \overline{\mathcal{D}(Q_j)}$ and let P_j be the orthogonal projection onto $\mathcal{H}_j$. Then for each $t > 0$

$$[(e^{tA_1/n} P_1)(e^{tA_2/n} P_2)]^n \to e^{tA_3} P_3$$

strongly on $\mathcal{H}$.

Given Theorems B and C, the proof of Theorem 1.1 is very easy:

PROOF OF THEOREM 1.1: The statements

$$-\Delta + n 1_{D^c} \uparrow -G_1$$

$$-\Delta + n 1_{\overline{D}^c} \uparrow -G_2$$

follow straight from the definitions. The statement

$$-\Delta_{D_n} \uparrow -G_2$$

is almost as immediate. For if we temporarily write

$$-\Delta_{D_n} \uparrow -\tilde{G}_2,$$

it follows from the definitions that

$$\mathcal{D}(\sqrt{-\tilde{G}_2}) = \bigcap_{n=1}^{\infty} H_0^1(D_n).$$

Clearly

$$\bigcap_{n=1}^{\infty} H_0^1(D_n) \subset \tilde{H}_0^1(\overline{D})$$

The reverse inclusion can be seen as follows. Suppose $f \in \tilde{H}_0^1(\overline{D})$ and $\eta \in C_0^{\infty}(\mathbb{R}^d)$. Fix n. Clearly, we can approximate ηf by a sequence of functions in $C_0^{\infty}(D_n)$. We need only smooth with a sequence of approximate identities in the standard way. This shows that $\eta f \in H_0^1(D_n)$ for all $\eta \in C_0^{\infty}(\mathbb{R}^d)$. It is easy to take $\eta \uparrow 1$ and conclude that $f \in H_0^1(D_n)$. This gives

$$\bigcap_{n=1}^{\infty} H_0^1(D_n) = \tilde{H}_0^1(\overline{D})$$

and thus $G_2 = \tilde{G}_2$.

The statements

$$(e^{t\Delta/n}\, 1_D)^n \longrightarrow e^{G_1 t} P_1^0$$

$$(e^{t\Delta/n}\, 1_{\overline{D}})^n \longrightarrow e^{G_2 t} P_3^0$$

follow from Theorem C. We let $A_1 = \Delta$ in $\mathcal{H}_1 = L^2(\mathbb{R}^d)$ and $A_2 = 0$ in $\mathcal{H}_2 = L^2(D)$. Then the first of the above two statements follows immediately. The second follows similarly by letting $\mathcal{H}_2 = L^2(\overline{D})$.

■

Before proceeding to relate the Sobolev spaces $H_0^1(D)$, $\tilde{H}_0^1(D)$, $\tilde{H}_0^1(\overline{D})$, we need some notation and definitions. For any Borel set $M \subset \mathbb{R}^d$, introduce the stopping time

$$\gamma_M = \inf \{t > 0: \int_0^t 1_{M^c}(X_s)ds > 0\}$$

where $\{X_t\}_{t\geq 0}$ is d-dimensional Brownian motion. (We also write $X_t = X(t)$.) Stroock has called γ_M the "first penetrating time" into M^c [11]. The stopping time γ_M should be compared to the first exit time from M,

$$\tau_M = \inf \{t > 0: X_t \in M^c\}.$$

Clearly,

$$\tau_M \leq \gamma_M$$

with equality if M is closed.

Put

$$M^* = \{x \in \mathbb{R}^d: P^x(\gamma_M > 0) = 1\}.$$

Ciesielski [3] has shown that $P^x(\tau_{M^*} = \gamma_M) = 1$, but we will not make use of this fact.

It is easy to see that

$$M^0 \subset M^* \subset \overline{M}.$$

The following definition extends slightly a definition given by Stroock [11].

<u>**DEFINITION**</u>: If D is open, D^c is called Kac-regular if for each $x \in D$

$$P^x\ (\tau_D < \infty,\ X(\tau_D) \in D^*) = 0. \tag{2.3}$$

The Blumenthal 0-1 law allows us to restate the definition of M^* as

$$P^x(\gamma_M > 0) = 1_{M^*}(x). \tag{2.4}$$

By the strong Markov property and (2.4),

$$P^x(\tau_D < \gamma_D) = E^x\left(\tau_D < \infty,\ P^{X(\tau_D)}(\gamma_D > 0)\right)$$

$$= P^x(\tau_D < \infty,\ X(\tau_D) \in D^*).$$

Hence we learn that D^c is Kac-regular if and only if

$$P^x\ (\tau_D = \gamma_D) = 1. \tag{2.5}$$

For our purposes, the importance of the notion of Kac-regularity is the following result.

THEOREM 2.1: *Suppose* D *is an open set in* $\mathbb{R}^d$. *Then*

$$H_0^1(D) = \tilde{H}_0^1(D)$$

if and only if D^c *is Kac-regular.*

We remark that according to Stroock ([11] Theorem 8), D^c is Kac-regular if and only if $D^* \cap \partial D$ is polar.

Our approach to proving Theorem 2.1 is to compare $e^{t\Delta_D}$ and e^{tG_1}. Since a similar approach will be used with $\tilde{H}_0^1(\overline{D})$, we need to examine e^{tG_2} as well. We thus construct the semigroup corresponding to the Sobolev space $\tilde{H}_0^1(M)$.

Let G_M be the self-adjoint operator corresponding to the form

$$Q(u,v) = (u,v)_*, \quad \mathcal{D}(Q) = \tilde{H}_0^1(M).$$

THEOREM 2.2. *For any Borel set* $M \subset \mathbb{R}^d$, *the closure of* $\tilde{H}_0^1(M)$ *is* $L^2(M^*)$. *For any* $f \in L^2(\mathbb{R}^d)$ *and* $t > 0$

$$e^{tG_M} 1_{M^*} f(x) = E^x(f(X_t);\ \gamma_M \geq t), \quad m\text{-a.e.} \tag{2.6}$$

PROOF. Let P_M be the orthogonal projection onto the closure of $\tilde{H}_0^1(M)$. We easily see that as $n \to \infty$

$$-\Delta + n 1_{M^c} \uparrow -G_M$$

so that by Theorem B, for $t > 0$

$$e^{t(\Delta - n 1_{M^c})} \longrightarrow e^{tG_M} P_M \tag{2.7}$$

strongly on $L^2(\mathbb{R}^d)$. By the Feynman-Kac formula

$$e^{t(\Delta - n 1_{M^c})} f(x) = E^x\, e^{-n \int_0^t 1_{M^c}(X_s)ds} f(X_t) \tag{2.8}$$

Since

$$\int_0^t 1_{M^c}(X_s)ds = 0 \Longleftrightarrow \gamma_M \geq t$$

we have for each Brownian path

$$e^{-n \int_0^t 1_{M^c}(X_s)ds} \longrightarrow 1_{\{t \leq \gamma_M\}}.$$

Thus, for each $x \in \mathbb{R}^d$, the right side of (2.8) converges to $E^x(f(X_t);\ \gamma_M \geq t)$. Since (2.7) holds for $t > 0$, we conclude that for $t > 0$

$$e^{tG_M}P_M f(x) = E^x(f(X_t);\ \gamma_M \geq t) \text{ m-a.e.} \tag{2.9}$$

Taking $t \downarrow 0$ in (2.9) with $f \in C_0^\infty(\mathbb{R}^d)$ gives

$$P_M f(x) = f(x)\ P^x(\gamma_M > 0) = 1_{M^*}(x) f(x) \text{ m-a.e.}$$

This gives the result. ■

The following interesting corollary is known (see Ciesielski [3]).

<u>COROLLARY 2.3</u>. Suppose $M \subset \mathbb{R}^d$ is a Borel set. Then

$$m(M^* \backslash M) = 0. \tag{2.10}$$

<u>PROOF</u>. The closure of $\tilde{H}_0^1(M)$ is $L^2(M^*)$ by Theorem 2.2. But by definition $\tilde{H}_0^1(M) = H^1(\mathbb{R}^d) \cap L^2(M)$ so that $L^2(M^*) \subset L^2(M)$. This is equivalent to (2.10). ■

The above corollary generalizes a well-known result. If we set

$$M^{\circledast} = \{x \in \mathbb{R}^d:\ P^x(\tau_M > 0) = 1\}.$$

Then, since $\tau_M \leq \gamma_M$, it follows that $M^{\circledast} \subset M^*$. The corollary thus gives

$$m(M^{\circledast} \backslash M) = 0.$$

Actually, the set $M^{\circledast}\backslash M$ is polar [8], a property not generally shared by $M^{*}\backslash M$.

We now give the

PROOF OF THEOREM 2.1. First note that

$$C_0^{\infty}(D) \subset H_0^1(D) \subset \tilde{H}_0^1(D) \subset L^2(D)$$

so that the closures of both $H_0^1(D)$ and $\tilde{H}_0^1(D)$ are equal to $L^2(D)$. We have

$$H_0^1(D) = \tilde{H}_0^1(D) \Leftrightarrow \Delta_D = G_D \Leftrightarrow e^{t\Delta_D} = e^{tG_D}$$

for all $t > 0$. By Theorem 2.2, for each $f \in L^2(D)$ and $t > 0$

$$e^{tG_D}f(x) = E^x(f(X_t);\ \gamma_D \geq t) \quad m\text{-a.e.}$$

If D^c is Kac-regular, then $P^x(\tau_D = \gamma_D) = 1$ for each $x \in D$ so that

$$e^{tG_D}f(x) = E^x(f(X_t);\ \tau_D \geq t) \ m\text{-a.e.} \tag{2.11}$$

But the right side of (2.11) is just $e^{t\Delta_D}f(x)$ ([2],[5],[9]). Thus $e^{t\Delta_D} = e^{tG_D}$ for all $t > 0$. Conversely, if $e^{t\Delta_D} = e^{tG_D}$ for all $t > 0$, then for any $f \in L^2(\mathbb{R}^d)$

$$0 = E^x(f(X_t);\ \gamma_D \geq t) - E^x(f(X_t);\ \tau_D \geq t)$$

$$= E^x(f(X_t);\ \tau_D < t \leq \gamma_D) \quad m\text{-a.e.}$$

Taking $f = 1_{\{|x|<n\}}$ and letting $n \to \infty$ gives

$$P^x(\tau_D < t \leq \gamma_D) = 0 \quad \text{m-a.e.}$$

for all $t > 0$. Since

$$\{\tau_D < \gamma_D\} = \bigcup_{\substack{t>0,\\ \text{rational}}} \{\tau_D < t \leq \gamma_D\},$$

we conclude that

$$P^x(\tau_D < \gamma_D) = 0 \quad \text{m-a.e.} \tag{2.12}$$

It follows that (2.12) holds for each $x \in D$ since by a standard argument [8], this function is harmonic in D. We conclude that $P^x(\tau_D = \gamma_D) = 1$ for each $x \in D$ which implies D^c is Kac-regular.

■

The following "uniqueness" theorem is an application of Theorem 2.1.

__THEOREM 2.3__. Let $D \subset \mathbb{R}^d$ be open with D^c Kac-regular and suppose that $u_i \in H^1(\mathbb{R}^d)$, $i = 1,2$ with $u_1 = u_2$ m-a.e. in D^c. If $\Delta u_i = 0$ in D, $i = 1,2$, then we have $u_1 = u_2$ m-a.e. in $\mathbb{R}^d$.

__PROOF__. Put $u = u_1 - u_2$. Then $u \in \tilde{H}_0^1(D)$. By Theorem 2.1 we have $u \in H_0^1(D)$. Since $\Delta u = 0$ in D, we have

$$\int \nabla\Phi \cdot \nabla u \, dx = 0$$

for all $\varphi \in C_0^\infty(D)$. Since we can approximate u by such φ in the $H^1(\mathbb{R}^d)$ norm, we conclude that

$$\int |\nabla u|^2 dx = 0.$$

This, coupled with the fact that $u \in H_0^1(D)$, leads to the conclusion $u = 0$ m-a.e. ■

We now state our theorem characterizing the open sets for which $\tilde{H}_0^1(D) = \tilde{H}_0^1(\overline{D})$.

<u>**THEOREM 2.4**</u>. Suppose $D \subset \mathbb{R}^d$ is open. Then $\tilde{H}_0^1(D) = \tilde{H}_0^1(\overline{D})$ if and only if

$$m(\overline{D}^* \cap \partial D) =$$

$$m(\{x \in \partial D: P^x(\tau_{\overline{D}} > 0) = 1\}) = 0.$$

<u>**REMARK**</u>. At first glance, one might guess that $\tilde{H}_0^1(D) = \tilde{H}_0^1(\overline{D}) \Leftrightarrow m(\partial D) = 0$. This is presumably not the case.

Theorem 2.4 follows easily from the following result concerning arbitrary Borel sets.

<u>**THEOREM 2.5**</u>. *Suppose M, M_1, M_2 are Borel subsets of $\mathbb{R}^d$. Then*

$$\tilde{H}_0^1(M) = \tilde{H}_0^1(M^*). \tag{2.13}$$

The equality $\tilde{H}_0^1(M_1) = \tilde{H}_0^1(M_2)$ holds if and only if

$$m(M_1^* \backslash M_2^*) + m(M_2^* \backslash M_1^*) = 0. \tag{2.14}$$

__PROOF__. By Theorem 2.2, $\tilde{H}_0^1(M) \subset L^2(M^*)$. Thus

$$\tilde{H}_0^1(M) = \tilde{H}_0^1(M) \cap L^2(M^*) = H^1(\mathbb{R}^d) \cap L^2(M) \cap L^2(M^*)$$

$$= H^1(\mathbb{R}^d) \cap L^2(M^*) = \tilde{H}_0^1(M^*).$$

Here we have used $L^2(M^*) \subset L^2(M)$ (see Corollary 2.3).

If $\tilde{H}_0^1(M_1) = \tilde{H}_0^1(M_2)$, then their closures, $L^2(M_1^*)$ and $L^2(M_2^*)$ are equal. This is equivalent to (2.14). Conversely, if $L^2(M_1^*) = L^2(M_2^*)$, then by (2.13)

$$\tilde{H}_0^1(M_1) = \tilde{H}_0^1(M_1^*) = H^1(\mathbb{R}^d) \cap L^2(M_1^*)$$

$$= H^1(\mathbb{R}^d) \cap L^2(M_2^*) = \tilde{H}_0^1(M_2^*) = \tilde{H}_0^1(M_2). \qquad \blacksquare$$

We can now give the proof of Theorem 2.4:

Since $\overline{D} \supset D$, we have $\overline{D}^* \supset D^*$ and thus $\tilde{H}_0^1(D) = \tilde{H}_0^1(\overline{D})$ if and only if

$$m(\overline{D}^* \backslash D^*) = 0. \tag{2.15}$$

But since D is open, $D^* \supset D$ while from Corollary 2.3 $m(D^* \backslash D) = 0$. Hence, (2.15) is equivalent to

$$m(\overline{D}^* \backslash D) = 0. \tag{2.16}$$

Since $\overline{D}$ is closed, $\overline{D}^* \subset \overline{D}$ so that (2.16) is equivalent to

$$m(\overline{D}^* \cap \partial D) = 0. \tag{2.17}$$

As we mentioned earlier, if M is closed, then $\tau_M = \gamma_M$. Thus $\overline{D}^* = \overline{D}^{\ast}$ and the result follows. $\blacksquare$

In our next result, we show that a well-known geometrical restriction on D suffices for the equality of $H_0^1(D)$ and $\tilde{H}_0^1(\overline{D})$.

DEFINITION (see [1]). An open set D in $\mathbb{R}^d$ satisfies the segment condition if for any $x \in \partial D$ there is a nonzero vector z_x and an open set U_x containing x such that for any $y \in \overline{D} \cap U_x$ and $t \in (0,1)$ we have $y+tz_x \in D$.

THEOREM 2.6. *Suppose* $D \subset \mathbb{R}^d$ *is open. If* D *satisfies the segment condition, then*

$$H_0^1(D) = \tilde{H}_0^1(\overline{D}).$$

PROOF. Suppose $f \in \tilde{H}_0^1(\overline{D})$. We must show $f \in H_0^1(D)$. We can assume $f(x) = 0$ for $|x| \geq R$ (some $R < \infty$) since if $\varphi \in C_0^\infty(\mathbb{R}^d)$ with $\varphi(x) = 1$ for $|x| \leq 1$, then setting $\varphi_n(x) = \varphi(x/n)$, we have

$$\varphi_n f \to f \quad \text{in } H^1(\mathbb{R}^d).$$

$(\nabla(\varphi_n f)(x) = n^{-1}\nabla\varphi(x/n)f(x)+\varphi_n(x)\nabla f(x) \xrightarrow{L^2} \nabla f(x).)$ Hence it suffices to show $\varphi_n f \in H_0^1(D)$ for each n. Thus we can assume $f(x) = 0$ for $|x| \geq R$. Let $B_t = \{x \in \mathbb{R}^d: |x| < t\}$.

Given $x \in \partial D$, let U_x and z_x be as in the definition above. A finite family $\{U_{x_i} : i = 1,2,\dots,N\}$ covers $\partial D \cap \overline{B}_{R+1}$. There exists a $\delta \in (0,1)$ such that if $\operatorname{dist}(x,\partial D \cap \overline{B}_{R+1}) \leq \delta$, then $x \in U_{x_i}$ for some i. Let

$$U_0 = D \cap B_{R+1} \cap \{x: \operatorname{dist}(x,\partial D \cap \overline{B}_{R+1}) > \delta\}.$$

Then the family $\{U_0, U_{x_1}, \ldots, U_{x_N}\}$ covers $\overline{D} \cap \overline{B}_R$. Let $\{\eta_i : i = 0,1,\ldots,N\}$ be a C_0^∞ partition of unity subordinate to this cover with supp $\eta_0 \subset U_0$ and supp $\eta_i \subset U_x$.

Note that $\eta_0 f \in H^1(\mathbb{R}^d)$ and

$$\eta_0 f(x) = 0 \quad \text{if } \operatorname{dist}(x,\partial D) \leq \delta.$$

If $\varphi \in C_0^\infty(\mathbb{R}^d)$ with $0 \leq \varphi$ and $\int \varphi(x)dx = 1$, let $\gamma_n(x) = \varphi(nx)n^d$. Then

$$\gamma_n * (\eta_0 f) \in C_0^\infty(D) \quad \text{for large } n$$

and $\gamma_n * (\eta_0 f) \to \eta_0 f$ in $H^1(\mathbb{R}^d)$. Thus $\eta_0 f \in H_0^1(D)$. It remains to show that

$$f_i = \eta_i f \in H_0^1(D),\ i = 1,\ldots,N.$$

By changing f_i on a set of measure zero, we can assume that

$$S_i \equiv \operatorname{supp} f_i \subset U_{x_i} \cap \overline{D}.$$

Let $f_i^t(x) = f_i(x - tz_{x_i})$. Then

$$\operatorname{supp} f_i^t = S_i + tz_{x_i}.$$

If $0 < t < 1$, $S_i + tz_{x_i} \subset D$. Since S_i is compact, $\operatorname{dist}(S_i + tz_{x_i}, \partial D) > 0$. Thus

$$\gamma_n * f_i^t \in C_0^\infty(D) \quad \text{for large } n.$$

Taking $n \to \infty$ gives $f_i^t \in H_0^1(D)$ for $0 < t < 1$. Taking $t \downarrow 0$, we obtain $f_i \in H_0^1(D)$. ∎

We end with an application of Theorems C and 2.2, which we found interesting.

__THEOREM 2.7__. *Let* M *be a Borel set in* $\mathbb{R}^d$. *Then for each* $t > 0$, *we have*

(i) $$P^x(X(kt/n) \in M,\ k = 1,\dots,n) \to P^x(t \le \gamma_M)$$

as $n \to \infty$ *in* $L^2_{loc}(\mathbb{R}^d)$,

(ii) *for* m-a.e. $x \in \mathbb{R}^d$ *and for all* $x \in M^0 \cup \overline{M}^c$

$$P^x(X(\frac{kt}{2^m}) \in M,\quad m = 1,\dots,\quad k = 1,2,\dots,2^m)$$

$$= P^x(t \le \gamma_M) = P^x(\int_0^t 1_{M^c}(X_s)ds = 0). \tag{2.18}$$

Here $f_n \to f$ *in* $L^2_{loc}(\mathbb{R}^d)$ *means that* $\int_K |f_n - f|^2 dx \to 0$ *for each compact set* $K \subset \mathbb{R}^d$.

__PROOF__. An easy application of Theorem C (as in the proof of Theorem 1.1) shows that if $t > 0$

$$(e^{t\Delta/n}1_M)^n \to e^{tG_M} 1_{M^*} \tag{2.19}$$

strongly on $L^2(\mathbb{R}^d)$. Using the Markov property and Theorem 2.2, we thus have for $f \in L^2(\mathbb{R}^d)$

$$E^x(f(X_t);\ X(tk/n) \in M,\ k = 1,2,\dots,n)$$

$$\xrightarrow{L^2(\mathbb{R}^d)} E^x(f(X_t);\ \gamma_M \ge t). \tag{2.20}$$

Given a compact set $K \in \mathbb{R}^d$ and $\varepsilon > 0$, choose N so that $P^x(|X_t| \geq N) \leq \varepsilon$. Then choosing $f = 1_{B_N}$, it easily follows from (2.20) that

$$\limsup_{n\to\infty} \int_K |P^x(X(tk/n) \in M, \quad k = 1,\ldots,n) - P^x(\gamma_M \geq t)|^2 dx$$

$$\leq \int_K 4\varepsilon^2 dx.$$

Hence, (i) is proved. If we let $n = 2^m$ and take $m\uparrow\infty$, $P^x(X(kt/n) \in M,\ k = 1,\ldots,n)$ decreases monotonically to a limiting value which from (i) is equal m-a.e. to $P^x\ (t \leq \gamma_M)$. This gives the equality in (2.18) m-a.e.

If $x \in \overline{M}^c$, then both sides of (2.18) are zero. We will show that both sides of (2.18) are continuous for $x \in M^0$. Abbreviate

$$Q = \{\frac{kt}{2^m}: \quad k = 1,2,\ldots,2^m, \quad m = 1,2,\ldots\}$$

and for $\delta \in [0,t)$, put

$$L_\delta(x) = E^x(P^{X(\delta)}(X(s-\delta) \in M,\ s \in Q,\ s > \delta))$$

$$R_\delta(x) = E^x(P^{X(\delta)}(t-\delta \leq \gamma_M))$$

Clearly, L_δ and R_δ are continuous for $\delta > 0$ and L_0 and R_0 are respectively the left and right sides of (2.18). Given $x_0 \in M^0$, there is an $r > 0$ so that $B_{2r}(x_0) = \{x: |x-x_0| < 2r\} \subset M^0$. We will show that

$$L_\delta \to L_0,\ R_\delta \to R_0$$

uniformly on $B_r(x_0)$ as $\delta\downarrow 0$.

For any $x \in B_r(x_0)$, we have by the Markov property

$$0 \le L_\delta(x)-L_0(x) \le P^x(X(s) \in M^c \text{ for some } s \in Q,\ s \le \delta$$

$$\le P^x(\tau_M \le \delta) \le P^0(\tau_{B_r} \le \delta) \downarrow 0$$

as $\delta \downarrow 0$.

Similarly,

$$0 \le R_\delta(x)-R(x) \le P^x\left(\int_0^\delta 1_{M^c}(X_s)ds > 0\right)$$

$$= P^x(\gamma_M < \delta) \le P^x(\tau_M < \delta) \le P^0(\tau_{B_r} < \delta) \downarrow 0$$

as $\delta \downarrow 0$. Hence, L_0 and R_0 are continuous on M^0. ■

References

1. R. Adams: Sobolev Spaces, Academic Press, New York, 1975.

2. K. Chung and Z. Zhao: "From Brownian Motion to the Schrödinger Equation," to appear.

3. Z. Ciesielski: "Lectures on Brownian Motion, Heat Conduction and Potential Theory," Mathematics Institute, Aarhus University, 1966.

4. W. Faris: Self-adjoint Operators, Springer Verlag, Berlin, 1975.

5. M. Fukushima: "Dirichlet Forms and Markov Processes," North-Holland, 1980.

6. T. Kato: Perturbation Theory for Linear Operators, Springer Verlag, Berlin, 1966.

7. T. Kato and M. Masuda: Trotter's product formula for nonlinear semigroups generated by the subdifferentials of convex functionals, J. Math. Soc. Japan 30, 169-178 (1978).

8. S. Port and C. Stone: Brownian Motion and Potential Theory, Academic Press, New York, 1978.

9. B. Simon: Functional Integration and Quantum Physics, Academic Press, New York, 1979.

10. __________: A canonical decomposition for quadratic forms with application to monotone convergence theorems, J. Functional Analysis 28, 377-385 (1978).

11. D. Stroock: The Kac approach to potential theory: Part I, J. Mathematics and Mechanics, vol. 16, no. 8, 829-852 (1967).

Mathematics Department
University of Virginia
Charlottesville, VA 22903

*Partially supported by NSF Grant DMS 8602826.

ON INVERTIBILITY OF MARTINGALE TIME CHANGES

Frank B. Knight[1]

1. Introduction. Let $(\Omega, \mathcal{F}_t, P)$, $0 \leq t \leq \infty$, be a probability filtration containing all P-null sets in $\mathcal{F}$, with $\mathcal{F}_t = \mathcal{F}_{t+}$ (the usual conditions). We assume that $\mathcal{F}_0 \equiv (\phi, \Omega)$, where $\equiv$ means "up to P-null sets", and that $L^2(\Omega, \mathcal{F}, P)$ is separable. Now suppose that we have a collection $\{M_k, k < N + 1\}$, $N \leq \infty$, of right-continuous, locally orthogonal and square-integrable $\mathcal{F}_t$-martingales starting at 0, such that $\mathcal{F}_t \equiv \sigma\{M_k(s), s \leq t, k < N + 1\}$ for each t. Alternatively, given any such local martingales we could replace $\mathcal{F}_t$ by the generated filtration to obtain this last condition. Or, with given $(\mathcal{F}_t, P)$ it is quite easy to show that for some $N \leq \infty$ (not unique) such M_k can be found. However, in the present paper we make the stronger assumptions:

A. The local martingales M_k, $k < N + 1$, have continuous paths,

and

B. For $k < N + 1$, $\langle M_1 \rangle_t = \ldots = \langle M_k \rangle_t \overset{(def)}{=} \langle M \rangle_t \to \infty$,

where $\langle M_k \rangle_t$ is the "quadratic variation process" of M_k, i.e. that these are all equal and increase to $+\infty$.[2]

[1]Research supported in part by N.S.F. D.M.S. 86-02304.

[2]We do not assume the "martingale representation property" (see Theorem 1.4 below). This would make N unique by a result of [3]. But our problem here is to start with N given martingales under minimal assumptions.

The "time change" is $\tau(t) = \inf\{s : \langle M\rangle_s > t\}$, and it is well-known that the process $B_k(t) \overset{(def)}{=} M_k(\tau(t))$ is an N-dimensional Brownian motion relative to $\mathcal{F}_{\tau(t)}$, and that $M_k(t) = B_k(\langle M\rangle_t)$ (see for example [7]). We are ready to introduce

Definition 1.1. Let $\mathcal{G}_t$ denote the filtration $\sigma(B_k(s), s \le t, k < N + 1)$, augmented by the P-null sets in $\mathcal{F}_\infty$. The time change $\tau(t)$ is "invertible" if $\langle M\rangle_t \in \mathcal{G}_\infty$ for each t.

Remarks. In the case $N = 1$, this is equivalent to $M_1(t)$ being "pure" in the sense of Dubins and Schwartz [4]. For examples of impure martingales which nevertheless satisfy the martingale representation property (of Theorem 1.4) see [15] and [20].

For general N, it follows by an argument of A. O. Pittenger (see [8, Lemmas 1.2, 1.3] for a more general result) that invertibility is equivalent to $\langle M\rangle_t$ being a stopping time of the filtration $\mathcal{G}_s$, for each t. To deduce this from [8, loc sit], one must show that $\{B_k(\langle M\rangle_t + u) - B_k(\langle M\rangle_t), 0 \le u, k < N + 1\}$ for each t is an N-dimensional Brownian motion independent of $\sigma\{B_k(\langle M\rangle_t \wedge s), 0 \le s, k < N + 1\}$, where $\wedge$ denotes a minimum. But since $B_k(\langle M\rangle_t) = M_k(t)$, this follows by applying the time change to the martingale $(M_k(t + u) - M_k(t), \mathcal{F}_{t+u})$ and noting that, for each k and s, $B_k(\langle M\rangle_t \wedge s) = M_k(t \wedge \tau(s)) \in \mathcal{F}_t$. Having shown that $\langle M\rangle_t$ is a stopping time, it now follows that invertibility is equivalent to $\mathcal{F}_t = \mathcal{G}_{\langle M\rangle_t}$ for all t. Indeed, $\sigma\{B_k(\langle M\rangle_t \wedge s), 0 \le s, k < N + 1\} \equiv \mathcal{G}_{\langle M\rangle_t}$ by a theorem of Blumenthal and Getoor [2], so that $\mathcal{G}_{\langle M\rangle_t} \subset \mathcal{F}_t$. Conversely, it suffices to observe that $M_k(s) = B_k(\langle M\rangle_s) \in \mathcal{G}_{\langle M\rangle_t}$ for $s \leq t$, by a familiar property of stopping times.

The purpose of the present work is to investigate the question "when is τ invertible?" This has been done before, especially by Stroock and Yor [15], [16]. But the problem is still far from a final solution. The following definition will be useful.

<u>Definition</u> 1.2. The time change $\tau(t)$ is "locally invertible" if, for every stopping time T of $\mathcal{F}_t$, $\langle M\rangle_{t\wedge T}$ is a stopping time of $\mathcal{G}_t$.

<u>Theorem</u> 1.3. If $\tau(t)$ is invertible, then it is locally invertible (and conversely).

<u>Proof</u>. It suffices to show that $\langle M\rangle_T$ is a stopping time of $\mathcal{G}_t$. But $\{\langle M\rangle_t \leq s\} = \{\tau(\langle M\rangle_t) \leq \tau(s)\}$, and since $\tau(\langle M\rangle_T) \geq T$ it is easy to show that $\tau(\langle M\rangle_T)$ is an $\mathcal{F}_t$ stopping time along with $\tau(s)$. Therefore, using $\mathcal{F}_t = \mathcal{G}_{\langle M\rangle_t}$, we have

$$\{\langle M\rangle_T \leq s\} \in \mathcal{F}_{\tau(s)} = \mathcal{G}_{\langle M\rangle_{\tau(s)}} = \mathcal{G}_s,$$

where the last equality follows because $\mathcal{F}_{\tau(s)} \subset \mathcal{G}_\infty$ and $\mathcal{F}_{\tau(s)}$ is independent of $(B_k(s + u) - B_k(s), 0 \leq u, k < N + 1)$. Indeed, this last remains an N-dimensional Brownian motion given $\mathcal{F}_{\tau(s)}$, since $(M_k(\tau(s) + u) - M_k(\tau(s)), \mathcal{F}_{\tau(s)+u})$ is a continuous martingale with orthogonal components.

In the present paper we are interested in determining if certain time changes are invertible. To do this, we provide a couple of general conditions, one necessary and one sufficient, for invertibility. We do not by any means have a method which will always decide the question, but perhaps the general problem is no longer as obscure as it seemed in [16] (see the end of its introduction). On the other hand, we have not been able to resolve the problem raised in [16] of whether $M(t) = \int_0^t B^{2n}(s)dB(s)$ is pure when $B(t)$ is Brownian motion and $n \geq 1$ is an integer. But we will comment briefly on it at the end.

<u>Theorem</u> 1.4. In order for $\tau(t)$ to be invertible, it is necessary that every square-integrable martingale $M(t)$ be continuous, and have a representation

$$M(t) = \sum_{k<N+1} \int_0^t h_k(s)dM_k(s), \tag{1}$$

with $\mathcal{F}_t$-predictable $h_k(t)$ (the martingale representation property).

<u>Proof</u>. Suppose that $\tau(t)$ is invertible. If every $M(t)$ is not continuous, then by the orthonalization procedure of Meyer [11] and Kunita-Watanabe [10], there exists a ("purely discontinuous") martingale $M(t) \not\equiv 0$ orthogonal to $M_k(t)$, $k < N + 1$. Then for some non-random $T > 0$, $M(t \wedge T)$ is non-constant (with positive probability) and it is orthogonal to each $M_k(t)$. Since $EM^2(T) < \infty$, by the optional sampling theorem $M(\tau(t) \wedge T)$ is square integrable,

orthogonal to $B_k(t)$ for all k, and $M(\tau(t) \wedge T) = E(M(T) \mid \mathcal{F}_{\tau(t)})$. But, as shown in the proof of Theorem 1.3, $\mathcal{F}_{\tau(t)} = \mathcal{G}_t$, so by a well-known theorem of K. Ito we have a representation $M(\tau(t) \wedge T) = \sum_{k<N+1}\int_0^t h_k(s)dB_k(s)$ for $\mathcal{G}_t$-predictable $h_k(s)$. This contradicts the orthogonality unless $M(\tau(t) \wedge T) \equiv 0$. But $\lim_{t\to\infty} EM^2(\tau(t) \wedge T) = EM^2(T) > 0$, so we have a contradiction. The same argument applies to any non-trivial $M(t)$ orthogonal to $M_k(\tau)$, $k < N + 1$, proving the martingale representation property.

We base our sufficient condition for invertibility on the Chacon-Jamison Theorem. Our detailed knowledge of this result is largely derived from [17], from which we have

Theorem 1.5. Let X_t be a homogeneous, strong-Markov process with r.c.ℓ.ℓ. paths and values in a Lusin space $(E,\mathcal{E})$. If there are no traps or holding points, and if, for a fixed initial probability μ on E, $A_t \in \mathcal{F}_t^\mu$ is a continuous, strictly increasing process tending to ∞, with continuous inverse $\tau(t)$, then $\mathcal{F}_\infty^\mu \equiv \sigma(X_{\tau(t)}, 0 \le t)$. Note: the augmentations are all for the minimal generated filtrations, relative to the usual P^μ. Actually, $A_t \in \mathcal{F}_\infty^\mu$ $(= \bigvee_t \mathcal{F}_t^\mu)$ suffices for the proof.

Proof. In the first place, it can be assumed without loss of generality that X_t is defined canonically on the space of all r.c.ℓ.ℓ. paths, as is easy to see (A_t may be defined as its right-limit along rational t, outside a fixed P^μ-null set). Then the σ-field $\mathcal{T}$ generated by the trajectories (invariant under time changes) is contained in $\sigma(X_{\tau(t)}, 0\le t)$ augmented by the P^μ-null sets. By Theorem 1.1 of [17], there is a regular conditional probability $P^{\mu}_{\Upsilon}(w,\cdot)$ over the coordinate σ-field $\mathcal{F}_{\infty}$, given Υ, such that $P^{\mu}\{P^{\mu}_{\Upsilon}(w,\cdot)$ is a unit mass at $p(w)\} = 1$. Moreover, by Theorem 2.1 (i) of [17], $P^{\mu}\{w = p(w)\} = 1$, i.e. the point mass is concentrated at w itself. It follows that, for each t, $f(X_t) = \int f(w(t))P^{\mu}_{\Upsilon}(X_{\tau(\cdot)},dw)$ for all $f \in b(\mathcal{E})$ (bounded, measurable on E), P^{μ}-a.s. Therefore, $X_{\tau(\cdot)}$ determines X_t, $0 \leq t$, up to P^{μ}-indistinguishability so that $X_t \in \sigma(X_{\tau(s)}, 0 \leq s)$ up to a P^{μ}-null set. The converse inclusion $X_{\tau(t)} \in \mathcal{F}^{\mu}_{\infty}$ is straightforward, so the proof is complete.

To apply this result to invertibility of time change, without seeking for generality we will list a sufficient set of conditions which illustrates the principle and applies to the examples we have been able to handle.

Theorem 1.6. Suppose that there is a Markov process X_t as in Theorem 1.5, with $P = P^{\mu}$ and $\mathcal{F}_t = \mathcal{F}^{\mu}_t$, where $(E,\mathcal{E}) = (R^n,\mathcal{B}^n)$ for some $n < \infty$ (the real Borel space), and suppose that there is an $(n \times N)$-matrix-valued measurable function $C(x_1,\ldots,x_n)$ for which

$$(1.1) \qquad X_t = X_0 + \int_0^t C(X_{s-})d\underline{M}_s, \; 0 \leq t,$$

where $\underline{M}_s$ is the column vector of components $(M_k(s), k < N + 1)$. Then, if the time-changed equation

$$(1.2)\qquad X_{\tau(t)} = \int_0^{\tau(t)} C(X_{s-})d\underline{M}(s)$$

$$= \int_0^t C(X_{\tau(u-)})d\underline{B}(u)$$

has $X_{\tau(t)}$ as a strict solution (i.e. measurable over $\mathcal{G}(t)$ for each t), the time change $\tau(t)$ is invertible.

Conversely, if (1.1) holds but (1.2) has no strict solution, then $\tau(t)$ is not invertible.

Proof. The converse is obvious, since, as seen in the proof of Theorem 1.3, $\mathcal{F}_{\tau(t)} = \mathcal{G}_t$ when $\tau(t)$ is invertible. On the other hand, if $X_{\tau(t)}$ is a strong solution of (1.2), then by Theorem 1.5 we have $\mathcal{F}_\infty \equiv \sigma(X_{\tau(t)}, 0 \leq t) \subset \mathcal{G}_\infty$, and so $\tau(t)$ is invertible.

Remark. In applying this result, the problem is to identify such a process X_t. In principle, this is not as difficult as it may appear at first glance. Indeed, if one chooses any r.c.ℓ.ℓ. process Y_t on some Lusin space $(E,\mathcal{E})$ such that $\sigma(Y_s, s \leq t) \equiv \mathcal{F}_t$ for all t (and, apart from $Y_t = \underline{M}(t)$, there are always such processes in the form of multidimensional martingales $\underline{Y}_t = (E(Y_n \mid \mathcal{F}_t))$ where (Y_n) is any basis for $L^2(\Omega,\mathcal{F},P)$) then one can introduce the prediction process Z_t of Y_t, which is the probability-valued, optional process $Z_t(S) = P(Y_{t+(\cdot)} \in S \mid \mathcal{F}_t)$, $S \in \mathcal{F}^o$(= the coordinate σ-field). Then Z_t is a homogeneous Markov process generating $\mathcal{F}_t$ in the required way [9], and the problem becomes that of finding a coordinate system for Z_t such that (1.1) and (1.2) hold. Of course, this is a difficult or impossible problem in general, but the above approach may be helpful in special cases.

2. Applications and Examples.

We discuss 4 types of example, of which the third--to decide when the martingale composition of an n-harmonic function with n-dimensional Brownian motion is invertible--seems of greatest intrinsic interest. For the first type, let S denote a closed, nowhere dense subset of $[0,1]$ having strictly positive Lebesgue measure. To be explicit, we will construct S by the Cantor method of removing first $(\frac{3}{8}, \frac{5}{8})$, then $(\frac{5}{32}, \frac{7}{32})$ and $(\frac{19}{32}, \frac{21}{32})$, then 4 more open intervals of length 4^{-3} about the midpoints of the 4 remaining intervals, and so forth. Thus the measure of all removed intervals is $\sum_{n=1}^{\infty} 2^{n-1}4^{-n} = \frac{1}{2}$, so that also $|S| = \frac{1}{2}$. Now let $W(s)$, $W(0) = 0$, be a standard (Wiener) Brownian motion, and consider

1) $$M_1(t) = \int_0^t 1_S(s)dW(s)$$

2) $$M_2(t) = \int_0^t 1_S(W(s))dW(s)$$

3) $$M_3(t) = \int_0^t 1 - 1_S(W(s))dW(s).$$

We will decide the invertibility of cases 1) - 3) with $N = 1$ and $\mathcal{F}_t$ the σ-field generated by $M_i(s)$, $s \leq t$, $1 \leq i \leq 3$ respectively. Case 1) is clearly invertible, because $\langle M\rangle_t = \int_0^t 1_S(s)ds$ is non-random. Hence $M_1(t) = B(\langle M\rangle_t) \in \mathcal{G}$, as required. The difficulty in case 2) is that we lack any very clear representation of the generated filtration (it is not even obvious to us that all of its martingales are continuous). However, we do know that $\langle M_2\rangle_t = \int_0^t 1_S(W(s))ds = \int 1_S(x)\ell(t,x)dx$, where $\ell(t,x)$ is the local time of W, and since $\langle M\rangle_t \in \mathcal{F}_t$ this is enough to show noninvertibility. Indeed, $\ell(t,x)$ is continuous in (t,x), and S has locally positive measure at each $x \in S$, whence it follows that the excursion intervals of W into S^c are precisely the level stretches of $\langle M_2\rangle_t$. Thus $\mathcal{F}_1$ contains events like {the first duration in S^c exceeding $\frac{1}{8}$ also exceeds $\frac{1}{4}$, and is complete by time 1}. This event, and the disjoint event {the first duration greater than $\frac{1}{8}$ is less than $\frac{1}{4}$, and complete by time 1}, both have positive probability, and both have up to a P-null set the same image of time-changed trajectories $B(t)$, $0 \leq t$, as we can easily see by splicing a duration of the opposite type into the path of $W(t)$ during one of its earlier excursions into S^c. Thus Case 2) is noninvertible.

Case 3) is also noninvertible, but it is very different from 2). We will argue that the generated filtration of 3) is the same as that of $W(t)$. Hence, since 2) and 3) are orthogonal martingales, the noninvertibility of 3) follows by Theorem 1.4. Noting that the filtration of 3) contains both the durations of excursions into S^c and the <u>increments</u> of $W(t)$ during the corresponding excursions, it is possible to identify the exact times of the successive up-crossings and down-crossings of $(\frac{3}{8}, \frac{5}{8})$. Then, we can fill in the successive up and down-crossings of the next two intervals $(\frac{5}{32}, \frac{7}{32})$ and $(\frac{19}{32}, \frac{21}{32})$ between the former, and so forth. Thus we see that the generated σ-field of M_3 contains all of the <u>crossings</u> of intervals in S^c. But then, for each t, one can go back to the most recent such up and down-crossings, to read off which (if any) of the intervals in S^c must contain the present value of $W(t)$. For example, after the first upcrossing of $(\frac{3}{8}, \frac{5}{8})$ we know that $W(t) > \frac{3}{8}$ until the first downcrossing, after which we know $W(t) < \frac{5}{8}$ until the next up-crossing, and so forth. Thus, at any time t, we know either that $W(t)$ is above the lower, or that $W(t)$ is below the upper endpoint, for each of the intervals of S^c. This either determines $W(t)$ uniquely, or else it determines the precise interval of S^c which contains $W(t)$ in its closure, as well as the precise time when $W(s)$ most recently reached the closure. Either that time is t and so $W(t)$ is determined, or else it is a $t_0 < t$, in which case $W(t)$ is determined by the increment $W(t) - W(t_0)$. In any case, we have $W(t) \in \mathcal{F}_t$, and the time change is noninvertible.

The second type of example is that of

$$M_t = \int_0^t (a1_{x\leq 0}(W_s) + b1_{x>0}(W_s))dW_s, \tag{2.1}$$

where $N = 1$ and $\mathcal{F}_t \equiv \sigma(M_s, s \leq t)$. We assume a and b are non-zero constants, $a \neq b$. Then $\langle M\rangle_t = \int_0^t (a^2 1_{x\leq 0}(W_s) + b^2 1_{x>0}(W_s))ds$ and if $a = -b$ then $a^{-1}M_t$ is already a Brownian motion, and the time change is trivially invertible. So we assume $a^2 \neq b^2$, in which case it is easy to see that $\mathcal{F}_t \equiv \sigma(W_s, s \leq t)$. Therefore, we can use W_t for Theorem 1.6. The equation (1.1) for W_t is

$$W_t = \int_0^t (a^{-1}1_{(x\leq 0)}(W_s) + b^{-1}1_{(x>0)}(W_s))dM_s. \tag{2.2}$$

which gives the time-changed equation

$$(2.3)\quad W_{\tau(t)} = \int_0^t a^{-1}1_{(x\leq 0)}(W_{\tau(s)}) + b^{-1}1_{(x>0)}(W_{\tau(s)})dB(s).$$

We will show that the time change is invertible if and only if a and b have the same sign. The sufficiency is a very special case of [16, Theorem (1.2)], but it seems worthwhile to illustrate our method by proceeding directly. Indeed, if a and b have the same sign, the invertibility follows immediately from Theorem 1.6 and the fact that (2.3) has a unique strong solution, according to the result of Nakao [12] for coefficients bounded below by $\varepsilon > 0$. From this, a well-known theorem of Yamada and S. Watanabe (see [19]) implies that the solution is strict, as required. (These references were also used for the proof in [16]).

The case where a and b have opposite signs hinges on a long forthcoming result of M. Barlow and E. Perkins [1], according to which (2.3) does not have a pathwise unique solution in this case. Indeed, there exist, relative to some filtered Brownian motion $(\Omega, \mathcal{F}_t, P, B_t)$, two pathwise distinguishable solutions. On the other hand, the solution of (2.3) is unique in law, i.e. weak uniqueness holds for (2.3). This is known for any bounded, measurable coefficient bounded away from 0 ([14, 7.3.3]). The noninvertibility now follows immediately from the last assertion of Theorem 1.6 and the following result, included with the consent of Ed Perkins.

<u>Lemma</u> 2.1. Let $\sigma(x)$ and $b(x)$ be bounded measurable functions, with $0 < \varepsilon < |\sigma|$. If the equation

$$X_t = x_0 + \int_0^t \sigma(X_s)dB_s + \int_0^t b(X_s)ds \tag{2.4}$$

has at most one solution in law (weak uniqueness), and if there exists any strict solution, then the solution is pathwise unique.

<u>Proof</u>. (E. Perkins). Let C_{x_0} denote the continuous functions of $t \geq 0$, starting at x_0. We wish to obtain a measurable function $\psi : C_{x_0} \to C_0$, such that for any solution X_t of (2.4) we have $B_{(\cdot)} = \psi(X_{(\cdot)})$a.s. From (2.4) it follows that $B_t = \int_0^t \sigma^{-1}(X_s)dY_s$, where $Y_t = X_t - x_0 - \int_0^t b(X_s)ds$ is a continuous, square-integrable martingale whose law does not depend on the choice of X_t. For $n \geq 1$, $w \in C_{x_0}$, let $\varphi_n(t,w) = 2^n\int_{(k-1)2^{-n}}^{k2^{-n}}\sigma^{-1}(w_s)ds$ for $k2^{-n} \leq t < (k+1)2^{-n}$, $k \geq 1$ and let $\varphi_n(t,w) = \sigma^{-1}(x_0)$ for $0 \leq t < 2^{-n}$. Then clearly the expression $\psi_n(t) = \int_0^t \varphi_n(s,X_s)dY_s$ defines a measurable function on C_{x_0}, since φ_n has step-function values and the integral reduces to a sum. On the other hand, since $|\varphi_n(t,w)|$ is bounded uniformly in n and w, and converges to $\sigma^{-1}(w(t))$ for a.e. t, it follows by dominated convergence that for any choice of solution X_t,

$$\lim_{n\to\infty} E\left(\int_0^t (\varphi_n(s,X_s) - \sigma^{-1}(X_s))dY_s\right)^2 = 0. \tag{2.5}$$

Passing to a subsequence n_k, whose choice depends only on the law of X_t, we can obtain a.s. uniform convergence in (2.5) on compact sets of t. Now we define $\psi(w)_t = \begin{cases} \lim_{k\to\infty} \psi_{n_k}(t) \\ 0 \end{cases}$

according as the former limit exists uniformly on compact sets of t, or not. Then it follows that for any solution X_t of (2.4), $B_{(\cdot)} = \psi(X_{(\cdot)})$a.s., as required.

To complete the proof of Lemma 2.1, let $X_1 = F(B)$ be any strict solution of (2.4), where F is a measurable functional $C_0 \to C_{x_0}$. Then we have the equivalence in law $X_1 \overset{d}{\equiv} X$, and therefore

$$(2.6) \qquad (X,B) \overset{d}{\equiv} (X_1,B).$$

But then $X_1 = F(B)$ implies also $X = F(B)$a.s., proving pathwise uniqueness.

The third type of example concerns the martingale composition of an analytic or harmonic function with a Wiener process [3]. In the former $(N = 2)$ case the invertibility always holds. This is presumably known to some (unidentified) specialists on Lévy's time change theorem, but the proof seems illustrative. In the harmonic case (where $N = 1$), invertibility is of course the exception rather than the rule. We will give a self-contained proof of the characterization in case $n = 2$, using complex variable techniques. For $n \geq 3$ the proof depends on a general theorem of differential geometry ([13]). We are grateful to Tom Salisbury for bringing this

---

[3]Part of this example was presented at the 1987 annual A.M.S. meeting in San Antonio, Texas, in the Special Session on Stochastic Processes and Analysis.

reference to our attention, which replaced a longer differential-geometric argument.

Let $f(z) = u(x,y) + iv(x,y)$ be a non-constant function analytic except for singularities forming a polar set in R^2. We assume that $(0,0) = \underline{0}$ is regular, and let $(W_1,W_2)_t = \underline{W}(t)$ be a Brownian motion on R^2, $\underline{W}(\underline{0}) = \underline{0}$. Let $\underline{M}(t) = (u(\underline{W}(t)), v(\underline{W}(t)))$. From the harmonicity of u and v it follows from Ito's formula ([6, Theorem 6]) that $\underline{M}(t)$ is an orthogonal pair of continuous local martingales, with

$$\langle u(\underline{W})\rangle_t = \langle v(\underline{W})\rangle_t = \int_0^t (u_x^2 + u_y^2)(\underline{W}(t))dt$$

tending to ∞ as $t \to \infty$ a.s., so our hypotheses are satisfied with $N = 2$.

Let $D_\varepsilon = \{(x_0,y_0) : f$ is analytic in the disk of center (x_0,y_0), radius $\varepsilon > 0\}$, and let $T_\varepsilon = \inf\{t : \underline{W}(t) \in D_\varepsilon^c\}$. Then $\lim_{\varepsilon\to 0} T_\varepsilon = \infty$ (since the singularities are polar), and it follows by applying Ito's formula on D_ε, then letting $\varepsilon \to 0$, that

$$\underline{M}(t) = \int_0^t U_s \cdot d\underline{W}(s), \tag{2.7}$$

where U_s denotes the matrix $\begin{bmatrix} u_x & u_y \\ -u_y & u_x \end{bmatrix}$ composed with $\underline{W}(s)$. Let us assume for the moment that $(u_x^2 + u_y^2)(0,0) \neq 0$. Then, since the zeros of $u_x^2 + u_y^2$ (i.e. of f') are at most finitely many in compact subsets of each D_ε, we can invert (2.7) to obtain

$$(2.8) \qquad \underline{W}(t) = \int_0^t U_s^{-1} \cdot d\underline{M}_s$$

$$= \int_0^t U_s^{*} \cdot (u_x^2 + u_y^2)^{-\frac{1}{2}} (\underline{W}(s))d\underline{B}^{*}_s ,$$

where U_s^* is the transpose of U_s and $\underline{B}^*(t) = \int_0^t (u_x^2 + u_y^2)^{-\frac{1}{2}} d\underline{M}(s)$ is again a Brownian motion on R^2. Indeed, this follows by substituting for $d\underline{M}_s$ from Ito's formula. Now it is well-known ([5, p. 164] plus localization) that (2.8) has a pathwise unique, and hence strict, solution for $\underline{W}$ in terms of $\underline{B}^*$, and hence in terms of $\underline{M}$, since $\underline{B}^*$ is $\sigma(\underline{M})$-measurable. Thus we have $\sigma(\underline{M}(s), s \leq t) \equiv \sigma(\underline{W}(s), s \leq t)$, and we can consider $\underline{W}(t)$ as the strong Markov process X_t generating $\mathcal{F}_t$ as in Theorem 1.6, with (2.8) in the role of (1.1). The time-changed equation is now

$$(2.9) \qquad \underline{W}(\tau(t)) = \int_0^t U_{\tau(s)}^{-1} \cdot d\underline{B}(s),$$

and arguing as before to avoid the zeros of $u_x^2 + u_y^2$ we conclude that this has a unique, hence strict, solution. Therefore the time change is invertible by Theorem 1.6, as long as $(u_x^2 + u_y^2)(0,0) \neq 0$.

To treat this last special case, a separate artifice seems to be needed. For $\varepsilon > 0$, the argument just given shows

$$\underline{W}(\tau(t)) - \underline{W}(\tau(\varepsilon)) = \int_\varepsilon^t U_{\tau(s)}^{-1} d\underline{B}(s) \text{ a.s.},$$

whence we see that

$$\tau(t) - \tau(\varepsilon) = \int_\varepsilon^t U_{\tau(s)}^{-2} ds.$$

Then letting $\varepsilon \to 0$, we also obtain (2.9) in this case, but the pathwise-uniqueness conditions of [5, loc sit] are no longer satisfied. But let $\underline{X}_t$ be any other solution adapted to $\mathcal{F}_{\tau(t)}$. Then $\underline{W}(\tau(t))$ circles (0,0) for $0 < t < \varepsilon$ in such a way that, for any $\varepsilon > 0$, $\underline{X}_t$ must intersect $\underline{W}(\tau(t))$ in $0 < t < \varepsilon$ with probability 1 (note that $\underline{X}_t$ spends no time at (0,0)). Then letting $T(\varepsilon) = \inf\{t > \varepsilon : \underline{X}(t) = \underline{W}(\tau(t))\}$, since $T(\varepsilon)$ is an $\mathcal{F}_{\tau(t)}$-stopping time and $\underline{W}(T(\varepsilon)) \neq 0$ a.s. on $\{T(\varepsilon) < \infty\}$, the strong Markov property of $(\underline{B}(t), \mathcal{F}_{\tau(t)})$ and the former uniqueness argument show that $X_t = \underline{W}(\tau(t))$ for $t \geq T(\varepsilon)$. Since $\lim_{\varepsilon \to 0} T(\varepsilon) = 0$ a.s., this proves the pathwise uniqueness in (2.9). Hence $\underline{W}(\tau(t))$ is again $\sigma(\underline{B}_{(\cdot)})$-measurable, and the time change is invertible.

We turn now to the invertibility of $u(\underline{W}(t))$ alone. More generally, we have the following

<u>Theorem</u> 2.2. Let $2 \leq n < \infty$, and let $u(x_1, \ldots, x_n)$ be non-constant harmonic on R^n except for a polar set of singularities, not including $\underline{0}$. Let $M(t) = u(\underline{W}(t))$, where $\underline{W}(t)$ is n-dimensional Brownian motion. Then the time change of $M(t)$ $(N = 1)$ is invertible if and only if u is either linear, or there is a hyperplane H of dimension $0 \leq m < n - 1$ such that u is a function of the distance to H. All of the invertible $M(t)$ are either scaled Brownian motions, or else Bessel diffusions written in the natural scale of Feller to become local martingales.

<u>Remark</u>. For $n - m \geq 3$, one has $\lim_{t\to\infty} \langle M \rangle_t < \infty$, contrary to assumption B, page 1. However, if we replace $B(t)$ by $B_0(t)$(=$B(t)$ absorbed at 0), $B_0(\infty) = 0$, the rest of Definition 1.1 applies.

Proof. We will first outline a proof in case $n = 2$ which does not carry over to $n > 2$ but is more elementary. If u is linear, $u(\underline{W}(t))$ is clearly Brownian, and the time change is invertible. In general, since $\langle \underline{M} \rangle_t = \int_0^t (u_x^2 + u_y^2)(\underline{W}(s))ds$ is in $\mathcal{F}_t$, the composition $w(\underline{M}_t)$, where $w(x,y) = \ell n|u_x - iu_y|$ is harmonic except at the zeros of ∇u or the singularities of u, is also in $\mathcal{F}_t$. The zeros of ∇u are at most countable since $u_x - iu_y$ is locally analytic and u is not constant. Thus if we start at $t = \varepsilon > 0$, $\underline{W}(\varepsilon)$ is a.s. not in the above set, and we have

$$(2.10) \qquad w(\underline{W}(t)) - w(\underline{W}(\varepsilon)) = \int_\varepsilon^t \nabla w(\underline{W}) \cdot d\underline{W}$$

$$u(\underline{W}(t)) - u(\underline{W}(\varepsilon)) = \int_\varepsilon^t \nabla u(\underline{W}) \cdot d\underline{W}.$$

Now these two equations can be inverted locally to write $\underline{W}(t) - \underline{W}(\varepsilon)$ in terms of $(dw(\underline{W}_s), du(\underline{W}_s))$, $\varepsilon < s < t$, provided that the Jacobian $(u_x w_y - u_y w_x)(\underline{W}_\varepsilon) \neq 0$, a.s. This is analogous to (2.8), and if (2.10) is locally invertible then the time change is noninvertible because the dimension is 2 instead of 1 (Theorems 1.3 and 1.4). Thus it remains to examine the set where $u_x w_y - u_y w_x = 0$. Unless $\nabla u = 0$ or $\nabla w = 0$, this implies that ∇u and ∇w are parallel, and hence that $(u_x - iu_y)(w_x - iw_y)^{-1}$ is real. The set $\nabla w = 0$ is also polar, because $\{w_x - iw_y = 0\}$ has no limit points within open sets where w is harmonic unless w is a constant, which would imply $|u_x - iu_y|$ constant, and hence that u is linear. On the other hand, if $\mathrm{Im}((u_x - iu_y)(w_x - iw_y)^{-1}) = 0$, since this ratio is locally analytic, then (x,y) must belong to a (locally) one-dimensional set by the inverse function theorem, or else the ratio reduces to a real constant. The former case defines a set which contains $\underline{W}_\varepsilon$ with probability 0, hence causes no difficulty.

Finally, if the ratio reduces to a real constant c, let (x_0, y_0) be a regular point for u, and let v be a local conjugate so that $f(z) = (u + iv)(z)$ is analytic near $z_0 = x_0 + iy_0$. Then $w_x - iw_y = \frac{d}{dz} \ell n\, f'(z)$ near z_0 (provided that $\nabla u(x_0, y_0) \neq 0$, as may be assumed), and so $(u_x - iu_y)(w_x - iw_y)^{-1} \equiv c$ (real) implies $f' = c \frac{d}{dz} \ell n\, f'$ near z_0. This equation is easy to solve, and the general solution gives $u = c_1\, \ell n|(x-x')^2 + (y-y')^2| + c_2$, real (c_1, c_2) and (x', y'). Conversely, for such u (with $(x', y') \neq (0,0)$), $u(\underline{W}(t))$ is a Bessel martingale ($\Rightarrow$ a regular one-dimensional diffusion) and the time change to $B(t)$ is well-known to be invertible in that case. This completes the argument for $n = 2$.

Turning to the case $n > 2$, we again can assume u nonlinear. The argument again hinges on the fact that $\langle M\rangle_t = \int_0^t \nabla \cdot \nabla u(\underline{W}(s))ds \in \mathcal{F}_t$, but, in general, there is no analog of the harmonic function w. However, we set $w(x_1, \ldots, x_n) = \nabla \cdot \nabla u$, and a straightforward calculation gives $\Delta w = 2\sum\sum\left[\frac{\partial^2 u}{\partial x_j \cdot \partial x_k}\right]^2 \geq 0$, since u is harmonic. This implies that $w(\underline{W}(t))$ is a continuous local submartingale, so we have a unique Doob-Meyer decomposition

$$w(\underline{W}(t)) = N(t) + A(t),$$

where $N(t)$ is an $\mathcal{F}_t$-local martingale and $A(t)$ is a non-decreasing, $\mathcal{F}_t$-measurable process, and both are continuous in t,

$A(0) = 0$. By Ito's Formula we have

$$dw(\underline{W}(t)) = \nabla w(\underline{W}(t)) \cdot d\underline{W}(t) + \frac{1}{2}\sum_{k=1}^{n} \frac{\partial^2 w}{\partial x_k^2}(\underline{W}(t))dt,$$

and therefore $\langle N\rangle_t = \int_0^t \nabla \cdot \nabla w(\underline{W}(t))dt$. Now if the time change is invertible, then by Theorem 1.4 we must have a representation $N(t) = \int_0^t h(s)dM(s)$, where $h(t)$ is $\mathcal{F}_t$-predictable, in such a way that

$$d\langle N\rangle_t = h^2(t)d\langle M\rangle_t$$

$$= h^2(t)\nabla \cdot \nabla u(\underline{W}(t))dt,$$

and therefore, if we have invertibility,

$$h^2(t) = \nabla \cdot \nabla w(\underline{W}(t))(\nabla \cdot \nabla u\underline{W}(t))^{-1} \tag{2.11}$$

for a.e.t, P-a.s.

The following lemma is no doubt true without (2.11), but we only require

<u>Lemma</u> 2.3. Under (2.11), for a.e.t there exist $\delta_n \to 0+$ such that P-a.s.

$$\int_t^{t+\delta_n} h(s)dM(s) = h(t)(M(t+\delta_n) - M(t)) + o(\delta_n^{\frac{1}{2}}).$$

<u>Proof</u>. Without changing the left-side, we may and shall replace

$h(s)$ by $(\operatorname{sgn} h(s))\left[|h(s)| \wedge \left|\frac{\nabla\cdot\nabla w(\underline{W}(s))}{\nabla\cdot\nabla u(\underline{W}(s)}\right|^{\frac{1}{2}}\right]$, and, since u is nonlinear and harmonic, $\nabla \cdot \nabla u(\underline{W}(s)) \neq 0$ a.s. for each $s > 0$. On the other hand, w is also real-analytic, so either $\nabla \cdot \nabla w \equiv 0$ and then w is a constant, or else $\nabla \cdot \nabla w(\underline{W}(s)) \neq 0$ a.s. for each $s > 0$. If $\nabla \cdot \nabla w \equiv 0$ the assertion is trivial by (2.11), so we assume the second case. For $K > 0$, let $T_K(t) = \inf\{s > t : (\nabla \cdot \nabla u(\underline{W}(s) \vee \nabla \cdot \nabla w(\underline{W}(s))) > K\}$, where $\vee$ denotes the maximum. Then T_K is a stopping time of $\mathcal{F}_t$, and letting $M_K(t) = M(t \wedge T_K)$ we have

$$(2.12)\quad E\left(\left(\int_t^{t+\delta} h(s)dM_K(s) - h(t)(M_K(t+\delta) - M_K(t))\right)^2 \mid \mathcal{F}_t\right)$$

$$= E\left(\int_t^{t+\delta} (h(s) - h(t))^2 d\langle M_K\rangle_s \mid \mathcal{F}_t\right)$$

$$\leq \delta K(1 + h(t))^2 .$$

Now since $h(s) = h(s,w)$ is measurable in (s,w), by a theorem of Lebesgue we have for $\varepsilon > 0$ and a.e.t,

$$\lim_{\delta\to 0+} \delta^{-1}\int_t^{t+\delta} 1_{|h(s)-h(t)|>\varepsilon} ds = 0, \text{ P-a.s.}$$

By the preceding argument, we have

$$\delta^{-1}\int_t^{t+\delta} (h(s)-h(t))^2 d\langle M_K\rangle_s$$

$$\leq K(1 + h(t))^2\delta^{-1}\int_t^{t+\delta} 1_{|h(s)-h(t)|>\varepsilon} ds + \varepsilon^2 K .$$

Since ε is arbitrarily small, it follows by dominated convergence that for a.e.t we have, P-a.s.

$$\lim_{\delta\to 0+} \delta^{-1}E\left(\int_t^{t+\delta} (h(s) - h(t))^2 d\langle M_K\rangle_s \mid \mathcal{F}_t\right) = 0.$$

But in view of (2.12) this implies that for a.e.t we can choose $\delta_n \to 0+$ such that

$$\text{(2.13)}\quad P\{\delta_n^{-1}\left(\int_t^{t+\delta_n} h(s)dM_K(s) - h(t)(M_K(t + \delta_n) - M_K(t))\right)^2 > 2^{-n}\} < 2^{-n} .$$

Since $\lim_{K\to\infty} T_K = \infty$ a.s., by passing to a subsequence (2.13) holds with $M(s)$ in place of $M_K(s)$, and Lemma 2.3 now follows by the Borel-Cantelli Lemma.

Next, since $w(\underline{W}(t + \delta)) - w(\underline{W}(t)) = \nabla w(\underline{W}(t)) \cdot (\underline{W}(t + \delta) - \underline{W}(t)) + o(\|\underline{W}(t + \delta) - \underline{W}(t)\|) \leq N(t + \delta) - N(t) = \int_t^{t+\delta} h(s)dM(s)$, it follows by Lemma 2.3 that for a.e.t

$$(\nabla w(\underline{W}(t)) \cdot (\underline{W}(t + \delta_n) - \underline{W}(t))\delta_n^{-\frac{1}{2}}(1 + o(1))$$

$$\leq \delta_n^{-\frac{1}{2}}h(t)(M(t + \delta_n) - M(t)) + o(1)$$

$$= h(t)\nabla u(\underline{W}(t)) \cdot (\underline{W}(t + \delta_n) - \underline{W}(t))\delta_n^{-\frac{1}{2}}(1 + o(1)) + o(1),$$

P-a.s. as $n \to \infty$. This implies that $\overline{\lim}_{n\to\infty} \delta_n^{-\frac{1}{2}}(\nabla w(\underline{W}(t)) - h(t)\nabla u(\underline{W}(t)))(\underline{W}(t + \delta_n) - \underline{W}(t)) \leq 0$, and using the fact that $\delta_n^{-\frac{1}{2}}(\underline{W}(t + \delta_n) - \underline{W}(t))$ is standard normal and independent of $(\underline{W}(t), h(t))$, it follows that in fact, for a.e.t, $\nabla w(\underline{W}(t)) - h(t)\nabla u(\underline{W}(t)) = 0$, P-a.s. Thus, for a.e.t, P-a.s. either $\nabla w(\underline{W}(t)) = \underline{0}$, or $\nabla w(\underline{W}(t))$ and $\nabla u(\underline{W}(t))$ are non-zero and parallel. But if $\nabla u(\underline{W}(t)) = \underline{0}$ with positive probability, then $\nabla w \equiv \underline{0}$ by analyticity (power series expansion), and so w is constant. In either case, the geometric meaning of the result is that the gradient of w is either 0 or parallel to the gradient of u, or in other words the level surfaces of u are also level surfaces of w, apart from the polar set of singularities. In particular, we conclude that if the time change is invertible, there is a point $\underline{x}_0 \in R^n$ and a neighborhood $V(\underline{x}_0)$ within which u is regular and for some fixed $g(x)$,

$$w(\underline{x}) = g(u(\underline{x})). \tag{2.14}$$

This will be enough to complete the proof by application of differential geometry, recalling that $w = \nabla \cdot \nabla u$.

More generally, suppose that u is twice continuously differentiable and satisfies equations of the form

$$\nabla \cdot \nabla u = g(u), \ \Delta u = h(u), \tag{2.15}$$

where g and h are measurable functions. According to [13, §2] it follows that the level surfaces of u are "a parallel family of hypersurfaces obtained from a hypersurface with constant principal curvatures." Furthermore, ([13, §3]) it was shown by B. Segre that the only hypersurfaces in R^{n+1} with constant principal curvatures, $2 \leq n$, are the hypercylinders of spherical dimension k, $0 \leq k \leq n$. If $k = 0$, we have a hyperplane (and u is linear) while for $k = n$ we have an n-sphere.

Therefore, arguing locally in $V(\underline{x}_0)$, and setting $u(\underline{x}_0) = c$, we find that $u(\underline{x})$ is a function of the distance from the hypercylinder $u(\underline{x}) = c$. Moreover, since $u(\underline{x})$ is analytic except for a polar set, this composition can be analytically extended wherever u is nonsingular. Finally, in our case u is also harmonic $(h(u) \equiv 0)$, and the only harmonic function of the distance is the Newtonian potential, namely (in dimension $k \geq 3$) $u(\underline{x}) = c_1(d)^{2-k} + c_2$ where c_1, c_2 are constants and d denotes the distance to the (n-k)-dimensional hyperplane of the centers. The easiest way to see this uniqueness is to note that our characterization makes $u(\underline{W}(t))$ a continuous function of a Bessel process of index k. Since u is harmonic, it is easy to see that this function cannot have a local maximum or minimum, hence it is monotone and $u(\underline{W}(t))$ is therefore a regular diffusion process. But it is also a local martingale, which implies that the Bessel process is in the natural scale of Feller. Now the natural scale is known to be unique up to linearity, so the expression for $u(\underline{x})$ obtained above is the most general.

Remark 1. The result of Theorem 2.2 is closely related to that of [17], which also relied on the same results from differential geometry. The problem of [17] was to determine all continuous

$h(x_1,\ldots,x_n)$ such that $h(\underline{W}(t) + \underline{x})$, for all $\underline{x} \in R^n$, is a Markov process. The main outcome was that $h = g(u)$ for a suitable continuous $g(x)$ and u as above, except that instead of $0 < m < n - 1$, one has $0 \leq m \leq n - 1$ since reflected Brownian motion is allowed but $|x_1|$ in R^n has a non-polar set of singularities. Why the answers to these two rather dissimilar problems should be so nearly the same is by no means clear.

Remark 2. The argument of [13] which we have relied on makes no pretense at completeness or comprehensibility to the non-specialist. The result itself is quite simple and appealing when understood properly, and we therefore outline our alternative proof. The details can be filled in, but they are purely technical. First, (2.14) shows that in $V(x_0)$ the magnitude of $\nabla \cdot u$ is constant on the level surfaces of u, which implies that these form a parallel family. Next, we choose an element of area in the level surface through x_0, and project it along the normals onto a parallel level surface at distance ε. If we fill in the normals on the boundary to obtain a closed (n-1)-dimensional surface, we can integrate the normal component of $\nabla \cdot u$ over the surface and obtain 0, since u is harmonic in $V(x_0)$. The contribution from the normals along the boundary is 0, and that of either of the level surfaces is proportional to the corresponding area, in inverse proportion to the corresponding magnitudes of $\nabla \cdot u$. Now the "main point" is that this ratio is constant over the surfaces, depending only on the separation $\pm \varepsilon$. When we write this out and let $\varepsilon \to 0$, it follows that each of the principal curvatures on each level surface is constant. The rest of the result then comes easily from the known characterization of such a surface, which also holds locally.

Finally, we would like to consider briefly an unsolved problem of [16, §2], which is to decide invertibility in the case $M(t) = \int_0^t W^n(s)dW(s)$ when $n \geq 2$ is even (if $n \geq 1$ is odd, $\tau(t)$ is invertible by [16]). It is easy to see that $\mathcal{F}_t \equiv \sigma\{W(s), s \leq t\}$ in this case, so we can try using $X_t = W(t)$ in Theorem 1.6. The time-changed equation can be written

$$\int_0^t W^n(\tau(s))dW(\tau(s)) = B(t). \tag{2.16}$$

Setting $B^*(t) = \int_0^t \operatorname{sgn} W(\tau(s))dB(s)$, which is also a Brownian motion, we have by time change in Tanaka's Formula

$$B^*(t) = \int_0^t W^n(\tau(s))d(|W(\tau(s))| - \ell(\tau(s))) \tag{2.17}$$

where $\ell(t)$ is the local time at 0 of $W(t)$. Now imitating the method of [16], let $Y(t) = |W(\tau(t)|^{n+1}$, and apply Ito's Formula to get (since $\ell(t)$ increases only when $W(t) = 0$)

$$Y(t) = (n+1)\int_0^t W^n(\tau(s))d(|W(\tau(s))| - \ell(\tau(s)))$$

$$+ \frac{1}{2}n(n+1)\int_0^t |W(\tau(s))|^{n-1}d\tau(s),$$

or again, since $d\tau(s) = W^{-2n}(\tau(s))ds$,

$$Y(t) - \frac{1}{2}n(n+1)\int_0^t Y^{-1}(s)ds = (n+1)B^*(t). \tag{2.18}$$

It follows as in [16] that this equation has a unique positive solution $Y(t)$ which is strict, i.e. $\sigma(B^*)$-measurable, but we do not know whether B^* is also $\sigma(B)$-measurable. However, a point not remarked in [16] is that this implies a determination of the law of $Y(t)$, and hence by symmetry that of $W(\tau(t))$. Indeed, it is easy to see from (2.18) that $(n+1)^{-1}Y(t)$ is a diffusion on $[0,\infty]$ with generator of the form $\frac{1}{2}\left(\frac{d^2}{dx^2} + \frac{n}{n-1} x^{-1} \frac{d}{dx}\right)$ at points $x > 0$. Since $Y(t)$ also has a zero set of Lebesgue measure 0, it can only be a reflected Bessel process of index $\frac{2n+1}{n+1}$. Then it is clear by symmetry that $(W(\tau(t))^{n+1}$ is a diffusion with the above generator for $x \neq 0$ and 0 a regular point having speed measure $m\{0\} = 0$, which determines its law uniquely.

The point of this digression here is that it leads to a remarkable (to us) nonuniqueness result. If we apply the same reasoning to (2.16) which led from (2.17) to (2.18), it follows (as also in [16]) that (2.18) holds with $(W(\tau(t)))^{n+1}$ in place of $Y(t)$ and $B(t)$ in place of $B^*(t)$. But since W^{n+1} is not positive, we do not have uniqueness. On the contrary, suppose we apply Ito's Formula, as extended by Meyer to differences of convex functions, to the function $(\operatorname{sgn} x)x^2$ composed with $(W(\tau(t)))^{n+1}$. Setting $(W(\tau(t)))^{n+1} = \gamma(t)$, and then $X_t = (\operatorname{sgn} \gamma(t))\gamma^2(t)$, we obtain by (2.18)

$$(2.19) \qquad X_t = 2\int_0^t (\operatorname{sgn} \gamma(s))\gamma(s)d\gamma(s) + \int_0^t \operatorname{sgn} \gamma(s)d\langle\gamma\rangle_s$$

$$= 2(n+1)\int_0^t |X_s|^{\frac{1}{2}}dB_s + (n+1)(2n+1)\int_0^t \operatorname{sgn} X_s ds.$$

Now (2.19) is known to have a positive solution $X^+(t)$, because if we replace $\operatorname{sgn} X_s$ by 1 it becomes the Ito equation for $Y^2(t)$ with B_s in place of B_s^* from (2.18). Moreover, this positive solution is known ([19]) to be pathwise unique. Hence X^+ is strict ($\sigma(B)$-measurable). But if we replace X^+ by $-X^+$ and dB by -dB (also Brownian) (2.19) is still satisfied, so we see that (2.19) (with dB) also has a strict negative solution $X^-(t)$. Besides these, there are many other strict solutions (as E. Perkins pointed out) obtained by switching from the positive solution to the negative one at stopping times such as $T = \inf\{t > 1 : X^+(t) = 0\}$. Of course, none of these is the solution X_t, which changes sign in the same way as the symmetric Bessel process of index < 2. The time change is invertible if (and only if) the process X_t is also $\sigma(B)$-measurable. But for this to be true would seem no more surprising than the fact that we already have such a diversity of strict solutons of (2.19).

REFERENCES

1. M. Barlow and E. Perkins, Sample path properties of stochastic integrals and stochastic differentiation. Preprint.

2. R. Blumenthal and R. Getoor, A theorem on stopping times, Annals of Math. Stat. 35(1964), 1348-1350.

3. M. Davis and P. Varaiya, The multiplicity of an increasing family of σ-fields, The Annals of Probability 2(1974), 958-963.

4. L. Dubins and G. Schwarz, On extremal martingale distributions, Proc. 5th Berkeley Symp. Math. Stat. Prob. II, Part 1(1967), 295-299.

5. N. Ikeda and S. Watanabe, Stochastic Diffferential Equations and Diffusion Processes, North-Holland (1981).

6. K. Ito, On a formula concerning stochastic differentials, Nagoya Math. J. 3(1951), 55-65.

7. F. Knight, On strict-sense forms of the Hida-Cramer representation, Seminar on Stochastic Processes, 1984, Birkhäuser (1986), 109-137.

8. F. Knight, Poisson representation of strict regular step filtrations, Sem. de Prob. XX, 1984/85, Springer Lecture Notes in Math. No. 1204(1986), 1-27.

9. F. Knight, Essays on the Prediction Process. Inst. of Math. Statist. Lecture Notes Series No. 1, (1981).

10. H. Kunita and S. Watanabe, On square-integrable martingales, Nagoya Math. J. 30(1967), 209-245.

11. P. A. Meyer, Probability and Potentials, Blaisdell (1966).

12. S. Nakao, On the pathwise uniqueness of solutions of one-dimensional stochastic differential equations, Osaka J. Math. 9(1972), 513-518.

13. K. Nomizu, Elie Cartan's work on isoparametric families of hypersurfaces, Proc. of Symposia in Pure Mathematics V27, Amer. Math. Soc. (1975), 191-200.

14. D. Stroock and S. Varadhan, Multidimensional Diffusion Processes, Springer-Verlag Grundlehren Series, Vol. 233(1979).

15. D. Stroock and M. Yor, On extremal solutions of martingale problems, Ann. Ecole Norm. Sup. 13(1980), 95-164.

16. D. Stroock and M. Yor, Some remarkable martingales, Sém. de Prob. XV, 1979/80, Springer Lecture Notes in Math. No. 850(1981), 590-603.

17. J. Walsh, On the Chacon-Jamison Theorem, Z. Wahrscheinlichkeitstheorie verw. Gebiéte 68(1984), 9-28.

18. A. Wang and C. Chen, Functions of an n-dimensional Brownian motion that are Markovian, Israel J. Math. 34(1979), 343-352.

19. T. Yamada and S. Watanabe, On the uniqueness of solutions of stochastic differential equations, J. of Math. of Kyoto Univ. 11(1971), 155-167.

20. M. Yor, Sur ℓ'étude des martingales continue extremales, Stochastics 2(1979), 191-196.

MULTIPLICATIVE MARTINGALES FOR SPATIAL BRANCHING PROCESSES.

by

J. NEVEU

Out of simplicity, we restrict ourselves to consider the dyadic brownian branching process $(N_t, t \in R_+)$ on the real line. By definition of this process, its particles perform independent brownian motions untill they split into exactly two particles at independent and mean one exponential times; then N_t denotes the point process formed on R by the particles alive at time t.

For each real x the formula

$$M_\phi^x(t) = \prod_{y \in N_t} \phi(x + y - \lambda t) \qquad (t \in R_+)$$

defines a martingale with values in $]0,1[$ provided the real function ϕ solves Kolmogorov's equation

$$\frac{1}{2}\phi'' - \lambda\phi' + \phi(1-\phi) = 0$$

on R and takes its values in $]0,1[$. Such a solution exists iff $\lambda^2 \geq 2$ and is then unique up to a translation; at the critical points $\lambda = \pm\sqrt{2}$, we show in paragraph 4 that these martingales are more informative than the usual additive martingales

$$W_a(t) = \sum_{y \in N_t} \exp(ay - v(a)t) \qquad (t \in R_+)$$

where $v(a) = 1 + a^2/2$, studied by Biggins [3] and Uchiyama [13].

In preparation to paragraph 4, we are recalling in paragraph 1 some results for the simple supercritical Galton-Watson process due to A. Joffe [11] which are mainly interesting when the reproduction law of the process is not of the L log L class; it will appear that the behavior of these "pathological" Galton-Watson processes is closely connected to the behavior of the spatial branching process in their critical directions.

In paragraph 2, dyadic brownian branching trees are introduced following the ideas developped in [12] and then short proofs of the convergence properties of the additive martingales above are included in paragraph 3 with some further results.

In paragraph 5 finally, we consider a first crossing problem for the particles of a dyadic brownian branching tree (trees are here unavoidable !). The numbers Z_s^λ of particles of this tree which cross the lines $x = \lambda t - s$ in the (x,t) - space $(s \in R_+)$ for the first time among their ancestors build a Galton - Watson process for which $(\phi(x-s)^{Z_s^\lambda}, s \in R_+)$ is a martingale whatever $x \in R$; this last property determines the law of the Galton - Watson process Z^λ uniquely as well as its asymptotic behavior. At the (most interesting) critical point $\lambda = \pm\sqrt{2}$, the Z^λ process is not of the L log L class and in fact $Z_s^\lambda / E(Z_s^\lambda)$ behaves as $1/s$ at infinity. We furthemore study the instants at which the Z_s^λ particles of the tree first - cross the lines $x = \lambda t - s$.

The first crossing problem studied in paragraph 5 is obviously the proper probabilistic setting for the Dirichlet problem attached to the operator $\frac{\partial}{\partial t} u + \frac{1}{2} \frac{\partial}{\partial x^2} u + u(1-u)$ relatively to the lines $x = \lambda t - s$. This will be developped elsewhere.

1. PRELIMINARIES ON THE GALTON-WATSON SUPERCRITICAL PROCESSES.

Consider a continuous time Galton-Watson process $(Z_s, s \in R_+)$ on N starting at $Z_o = 1$, with infinitesimal generating function

$$a(u) = \alpha \left[\sum_N p(k)\, u^k - u \right] \qquad (0 \le u \le 1)$$

where α is the death ratio of an individual and $p(\cdot)$ governs its reproduction. Assume that $0 < \mu < \infty$ where

$$\mu = \alpha \sum_N p(k)\,(k-1) ;$$

then $a(u) = 1$ has an unique solution σ in $[0,1[$ and $a(\cdot) < 0$ on $]\sigma,1[$. In particular for a process without extinction $(p(o) = 0)$, μ is necessarily > 0 outside the degenerate case $(p(1) = 1)$ and then $\sigma = 0$.

The equation

$$(1.1) \qquad \psi' = a \circ \psi \quad \text{on} \quad R$$

has a solution with values in $]\sigma,1[$ which is unique up to a translation. This function is of class C^∞, decreases strictly $(\psi' < 0)$, is such that $\psi(-\infty) = 1$, $\psi(+\infty) = \sigma$ and determines explicitedly the generating functions $f_s(u) = E(u^{Z_s})$ of the process by the relations

$$(1.2) \qquad f_s[\psi(u)] = \psi(u+s) \qquad (u \in R \;,\; s \in R_+)$$

At $-\infty$ more precisely

$$(1.3) \qquad \lim_{t \to -\infty} \log \psi(u+t) \,/\, \log \psi(t) = e^{\mu u} ;$$

the limit $\lim_{t \to -\infty} e^{-\mu t} \log \psi(t)$ allways exists but is different from zero if and only if $p(\cdot)$ belongs to the $L \log L$ class (i.e. if $\sum_{k \geq 2} p(k)\, k \log k < \infty$).

For every real s, $\psi(s-t)^{Z_t}, t \in R_+)$ is a martingale with values in $[0,1]$ which thus converges a.s. as $t \uparrow \infty$. For $s = 0$ this shows that

$$W = \lim_{t \to \infty} \text{a.s.} \ (-\log \psi(t))\, Z_t$$

exists and defines a random variable with values in $[0,\infty]$; hence by (1.2), the martingale $\psi(s-t)^{Z_t}$ goes a.s. to $\exp(-e^{\mu s} W)$ when $t \uparrow \infty$, for every $s \in R$ and since this convergence also holds in L^1, the Laplace transform of W is given by

$$(1.4) \qquad E\,[\exp(-e^{\mu s} W)] = \psi(s) \qquad (s \in R)\ .$$

In particular for $s \to \pm\infty$, one sees that $P(W = 0) = \psi(-\infty) = \sigma$ and that $P(W < \infty) = \psi(+\infty) = 1$. Finally the limit $\lim_{t \to \infty} Z_t\, e^{-\mu t}$ allways exists a.s. but is different from 0 (and then proportional to W) if and only if p belongs to the $L \log L$ class.

2. THE DYADIC BROWNIAN BRANCHING TREE.

Let us first define the canonical space Ω of the continuous trajectories of a branching system of particles moving on R, starting at time 0 at the origin; each particle is supposed to perform a continuous movement during its finite life time and to give birth to exactly two particles at its death time and its death location.

Finite sequence u of 1's and 2's will label the particles starting with the first particle labelled $\emptyset$; each particle u

has two descendants u1 and u2. Let $U = \sum_{N} \{1,2\}^n$ be the space of labels. Let τ_u be the lifetime of particle u $(u \in U)$ which will thus birth at time

$$T_u = \sum_{k=0}^{n-1} \tau_{j_1 \dots j_k} \quad \text{if} \quad u = j_1 j_2 \dots j_n \quad (u \in U) \; ;$$

the τ_u are assumed to be strictly positive reals satisfying the non-explosion condition : $\{u : T_u \le t\}$ is finite for every $t \in R_+$. The trajectories of particles are continuous maps X_u of the time intervals $[T_u , T_u + \tau_u]$ into R such that $X_\emptyset(0) = 0$ and $X_{uj}(T_{uj}) = X_u(T_u + \tau_u)$ for every $u \in U$ and $j = 1,2$.

A point ω of the canonical space Ω is a collection $(\tau_u , X_u ; u \in U)$ of reals and maps satisfying the above conditions. A family $(N(t,\omega) , t \in R_+)$ of finite point measures on R is associated to each $\omega \in \Omega$ by the formula

$$(2.1) \qquad N(t,\omega) = \sum_{u \in \mathcal{Z}(t,\omega)} \delta_{X_u(t)}$$

where $\mathcal{Z}(t,\omega) = \{u : T_u \le t < T_u + \tau_u\}$ is the set of particles alive at time t. Clearly $t \to N(t,\omega)$ is right continuous and left limited in t. Notice that the tree structure of ω is a much richer data then the process $(N(t) , t \in R_+)$; for instance the process $(N^{u,s}(t) , t \in R_+)$ of the descendants of u after time s $(u \in U , s \in R_+)$ can be defined for the ω such that $u \in \mathcal{Z}(s,\omega)$ by

$$(2.2) \qquad N^{u,s}(t) = \sum_{v : uv \in \mathcal{Z}(t+s)} \delta_{X_{uv}(s+t) - X_u(s)} \qquad (t \in R_+) \; .$$

but cannot be deduced from the $N(t)$'s ! Obviously

$$(2.3)\qquad N(s+t) = \sum_{u\in \mathcal{Z}(s)} \delta_{X_u(s)} * N^{u,s}(t)$$

A natural filtration $(\mathcal{F}_t , t\in R_+)$ on Ω is generated by $(\mathcal{Z}(t) , (X_u(t) , u\in \mathcal{Z}(t)))_{t\in R_+}$ and there exists an unique probability P on $(\Omega,\mathcal{F}_\infty)$ such that

$$(B_u(t) = X_u(T_u+t) - X_u(T_u) , 0\le t\le \tau_u)_{u\in U}$$

is an independent family of brownian motions starting at the origin and stopped after an exponential time of mean one. This probability fundamentally has the following branching property :

For every $s\in R_+$, $\mathcal{F}_s$ - conditionnally, the processes $N^{u,s}$ $(u\in \mathcal{Z}_s)$ are independent and follow the same law as N, i.e.

$$(2.4)\qquad E^{\mathcal{F}_s}[\prod_{u\in \mathcal{Z}_s} f_u(N^{u,s})] = \prod_{u\in \mathcal{Z}_s} \int f_u(N)\, dP .$$

The following easy formula will also be useful : as $t\downarrow 0$

$$(2.5)\qquad E[\int_R h(x)\, N(t,dx)] = h(0) + t[h(0) + \frac{1}{2} h''(0)] + o(t)$$

provided h is a bounded C^2 function on R.

3. ADDITIVE MARTINGALES.

For every real a, the t-function $E[\int e^{ax} N(t,dx)]$ is an exponential by (2.3) and (2.4), equal to $\exp[v(a)t]$ with $v(a) = 1+(a^2/2)$ by (2.5). Then by the branching property

$$(3.1)\qquad W_a(t) = \int_R \exp(ax - v(a)t)\, N(t,dx)$$

is a positive martingale on $(\Omega,\mathcal{F} ; \mathcal{F}_t , t\in R_+)$ belonging to L^q for every finite q as is easily checked. The following proposition concerning the a.s. limit of this martingale

$$W_a = \lim_{t\to\infty} \text{a.s. } W_a(t)$$

is due to Biggins [4] and Uchiyama [13]. We will indicate a shorter proof due to B. Chauvin and myself with some suplementary results.

Proposition 1.

a) For every $p \in]1,2]$, the martingale $(W_a(t), t \in R_+)$ is L^p-bounded provided $p\, a^2/2 \le 1$; hence this martingale is L^1-convergent provided $a^2/2 < 1$.

b) Allmost surely $W_a = 0$ when $a^2/2 \ge 1$.

Proof. 1) We rely on the following simple lemma which does not seem to be known (for $p \neq 2$!).

Lemma. Let $p \in]1,2]$. For any finite sequence of positive independent random variables $X_1, X_2, \ldots, X_n$ in L^p and any sequence of positive real numbers $c_1, \ldots c_n$

$$v_p\left(\sum_1^n c_j X_j\right) \le \sum_1^n c_j^p\, v_p(X_j)$$

where by definition $v_p(X) = E[X^p] - E[X]^p$.

To prove this lemma it obviously suffices to show that any couple of positive random variables X, Y such that $E[X/Y] = E(X)$ verify the inequality $v_p(X+Y) \le v_p(X) + v_p(Y)$. But by the concavity of the x function $(x+a)^p - x^p$ on R_+ for every $a > 0$ and for $p \in]1,2]$:

$$E[(X+Y)^p - X^p/Y] \le E[(X+Y)/Y]^p - E[X/Y]^p$$
$$= [E(X) + Y]^p - [EX]^p ;$$

hence

$$E[(X+Y)^p] - E[X^p] \le E[(EX+Y)^p] - [EX]^p .$$

Applying this formula to Y and EX in place of X and Y gives

$$E[(Y+EX)^p]-E(Y^p) \quad \le \quad [EY+EX]^p-[EX]^p$$

and combining the two last inequalities gives the desired result. □

The first part of the proposition now follows from the usual decomposition

$$(3.2) \qquad W_a(s+t) = \sum_{u \in \mathcal{Z}(s)} \exp[a\, X_u(s)-\phi(a)s]\; W_a^{u,s}(t)$$

where $W_a^{u,s}$ is defined from $N^{u,s}$ as W_a is from N, since by the lemma relativitezed to conditional expectations and the branching property :

$$E^{\mathcal{F}_s}[W_a(s+t)^p] \; - \; [W_a(s)]^p$$

$$\le \sum_{u \in \mathcal{Z}(s)} \exp(p[a\, X_u(s)-\phi(a)s]) \quad v_p[W_a(t)]$$

Taking expectations leads to

$$E[W_a(s+t)^p] \; - \; E[W_a(s)^p]$$

$$\le \; \exp([\phi(ap)-p\,\phi(a)]s) \quad v_p[W_a(t)]$$

for every $s \in R_+$; then the martingale $(W_a(t)\,,\, t \in R_+)$ is L_p-bounded if $\phi(ap) < p\,\phi(a)$, i.e. if $p\, a^2/2 < 1$.

2. Formula (3.2) for s equal to the first branching time $\tau := \tau_\phi$ and for $t = +\infty$ reads

$$W_a = \mathcal{Z} \cdot (W_a^{(1)} + W_a^{(2)})$$

$$\text{with} \quad \mathcal{Z} = \exp[a\, X_\emptyset(\tau) - \phi(a)\tau]$$

where $\mathcal{Z}$, $W_a^{(1)}$, $W_a^{(2)}$ are independent and $W_a^{(1)}$, $W_a^{(2)}$ have the same distribution as W_a. Then taking $p^{\underline{th}}$ moments for $p \in]0,1[$ and computing $E(\mathcal{Z}^p)$ leads to the equality

$$(3.3) \qquad E[(W_a^{(1)}+W_a^{(2)})^p] = (1+p\,\phi(a) - \tfrac{1}{2}p^2 a^2)\; E[W_a^p]\,.$$

But $E[(W_a^{(1)} + W_a^{(2)})^p] \le 2\, E[W_a^p]$ since $p < 1$ and $1 + p\,\phi(a) - \frac{1}{2} p^2 a^2 > 2$ if $p\, a^2/2 > 1$; this forces $W_a = 0$ a.s. when $a^2/2 > 1$ by a suitable choice of p.

To show that $W_a = 0$ a.s. even if $a^2/2 = 1$, let $p \uparrow 1$ in the above equality as follows. Notice first that for each $w_1 > 0$ and $w_2 \ge 0$

$$\frac{1}{1-p}\, w_1 [w_1^{p-1} - (w_1 + w_2)^{p-1}]$$

$$\begin{cases} \le \max(1, w_2) \\ \longrightarrow w_1 \log[(w_1 + w_2)/w_1] \quad \text{as} \quad p \uparrow 1\,; \end{cases}$$

Hence by dominated convergence

$$\lim_{p \uparrow 1} \frac{1}{1-p} (E[(W_a^{(1)})^p] - E[W_a^{(1)} (W_a^{(1)} + W_a^{(2)})^{p-1}])$$

$$= E\left[W_a^1 \log \frac{W_a^1 + W_a^2}{W_a^1}\right]$$

But by symmetry

$$E[W_a^{(1)} (W_a^{(1)} + W_a^{(2)})^{p-1}] = \frac{1}{2} E[(W_a^{(1)} + W_a^{(2)})^p]$$

so that the above formula gives at the limit

$$E\left[W_a^{(1)} \log \frac{W_a^{(1)} + W_a^{(2)}}{W_a^{(1)}}\right] = \frac{1}{2} (1 - \frac{a^2}{2})\, E(W_a)\,. \tag{3.4}$$

The proof is now completed since at the critical points $(a^2/2 = 1)$ the first member is zero which implies that $P(W_a^{(1)} > 0\,, W_a^{(2)} > 0)$ and then that $W_a = 0$ a.s. □

<u>Remark</u>. When $a^2/2 < 1$, the equality (3.4) just obtained easily implies that

$$P[W_a \geq x(1-\frac{a^2}{2})] \leq [x\ 2\log(2)]^{-1/2} \quad (x \in R_+)$$

thus showing that the laws of $W_a / (1-\frac{a^2}{2})$ are tight when a approaches the critical points. □

Finally let N_t^* be the position of the particle of $N(t)$ most to the right on R_+. Since $\sqrt{2}N_t^* - 2t \leq \log W_{\sqrt{2}}(t)$ it follows from the preceding theorem that

$$\text{(3.5)} \qquad \lim_{t\to\infty} \text{a.s. } N_t^* - \sqrt{2}t = -\infty .$$

In fact this can be proved much more simply and a much more refined result has been found by Bramson [5].

4. MULTIPLICATIVE MARTINGALES.

If $\phi : R \to]0,1[$ is a C^2 function, solution of Kolmogorov's equation

$$\text{(4.1)} \qquad \frac{1}{2}\phi'' - \lambda\phi' = \phi(1-\phi)$$

for a real λ, the next formula

$$M_\phi(t) = \prod_{u \in \mathcal{Z}(t)} \phi[X_u(t) - \lambda t]$$

$$= \exp\left[\int \log \phi(x-\lambda t)\, N(t,dx)\right]$$

defines a $\mathcal{F}_\bullet$ martingale $M_\phi = (M_\phi(t)\ ,\ t \in R_+)$ with values in $]0,1[$. This follows for instance from the semi-group theory developed in [10].

But (4.1) has a solution mapping the whole of R into $]0,1[$ if and only if $\lambda^2 \geq 2$ and then the solution ϕ is unique up to a translation of its argument; i.e. $(\phi(y+\cdot)\ ,\ y \in R)$ is the set of all solutions. Moreover $\phi' < 0$ on R and towards $-\infty$ where $\phi \to 1$:

$$(4.2) \qquad -\log \phi(x) \sim c\, e^{ax} \quad \text{if} \quad \lambda > \sqrt{2}$$
$$\sim (c' - cx)e^{ax} \quad \text{if} \quad \lambda = \sqrt{2}$$

where $c > 0$ and $c' \in R$, where $a = \lambda - \sqrt{\lambda^2 - 2}$ is $< \sqrt{2}$ if $\lambda > \sqrt{2}$ and is equal to $\sqrt{2}$ if $\lambda = \sqrt{2}$ (the proper relation between a and λ is $v(a) = a\lambda$); these asymptotic results imply that whatever $\lambda \geq \sqrt{2}$

$$(4.2') \qquad \lim_{x \to -\infty} \log \phi(y+x) / \log \phi(x) = e^{ay} \qquad (y \in R) .$$

As $t \to \infty$, the martingale M_ϕ converges a.s. and in L^1 to its limit that we will denote by $\exp(-\mathcal{Z}_\phi)$; the random variable $\mathcal{Z}_\phi$ then takes its values in $[0,\infty]$. Since $N_t^* - \lambda t \to -\infty$ a.s. when $t \to \infty$ and because $\lambda \geq \sqrt{2}$ by the end result of last paragraph, (4.2') shows that a.s.

$$\int_R \log \phi(y+x-\lambda t)\, N(t,dx) \Big/ \int_R \log \phi(x-\lambda t)\, N(t,dx) \longrightarrow e^{ay}$$

as $t \to +\infty$; hence $\mathcal{Z}_{\phi(y+\cdot)} = e^{ay}\, \mathcal{Z}_\phi$. The L^1-convergence of the martingales then implies that

$$M_{\phi(y+\cdot)}(t) = E^{\mathcal{F}_t}[\exp(-e^{ay}\, \mathcal{Z}_\phi)] \quad (y \in R\,,\ t \in R_+)$$

and in particular for $t = 0$, one obtains the Laplace transform of the variable $\mathcal{Z}_\phi$ in the form

$$(4.3) \qquad E[\exp(-e^{ay}\, \mathcal{Z}_\phi)] = \phi(y) \qquad (y \in R) .$$

This last formula implies that $0 < \mathcal{Z}_\phi < \infty$ a.s. since $\phi(-\infty) = 1$ and $\phi(+\infty) = 0$.

When $\lambda > \sqrt{2}$, it follows from (4.2) that

$$\lim_{t\to\infty} \int - \log \phi(x-\lambda t)\ N(t,dx) \Big/ \int \exp a(x-\lambda t)\ N(t,dx) = c \quad \text{a.s.}$$

for a constant $c > 0$, since $N_t^* - \lambda t \to -\infty$ a.s. when $t \to \infty$. Hence the a.s. convergence of the martingale $M_\phi(\cdot)$ is equivalent to that of the martingale $W_a(\cdot)$ and $Z_\phi = cW_a$. The situation is different and more interesting at the critical value $\lambda = \sqrt{2}$; here proposition 1 shows only that

$$\int \exp[\sqrt{2}(x - \sqrt{2}t)]\ N(t,dx) \equiv W_{\sqrt{2}}(t) \longrightarrow 0$$

a.s. when $t \to \infty$

whereas the consideration of the martingale M_ϕ leads to the following result.

<u>Proposition 2</u>. <u>The martingale defined by</u>

$$W'(t) = \int (\sqrt{2}t - x)\ \exp(\sqrt{2}(x - \sqrt{2}t))\ N(t,dx) \qquad (t \in R_+)$$

<u>converges a.s. when</u> $t \to \infty$ <u>to a random variable</u> W' <u>with values in</u> $]0,\infty[$ <u>and with infinite mean</u>. <u>Its (exponentially rescaled) Laplace transform</u>

$$\psi(y) = E[\exp(-e^{\sqrt{2}y}\ W')] \qquad (y \in R)$$

<u>is a solution</u> $\psi : R \to]0,1[$ <u>of Kolmogorov's equation</u> : $\frac{1}{2}\psi'' - \sqrt{2}\,\psi' = \phi(1-\phi)$.

<u>Proof</u>. Since $W'(t) = (-\frac{d}{da} W_a(t))_{a=\sqrt{2}}$, it is easy to prove that $W'(\cdot)$ is our integrable martingale. Then by (4.2), by the a.s. convergence of M_ϕ for $\lambda = \sqrt{2}$ and by the limiting behavior $\lim_{t\to\infty} N_t^* - \sqrt{2}t = -\infty$, this martingale a.s. converges to $\frac{1}{c} Z_\phi$. Hence $0 < W' < \infty$ and ψ coincides with ϕ up to a translation by $\frac{1}{\sqrt{2}} \log c$. Finally by (4.2) again

$$E(W') = \lim_{y \to -\infty} - \log \psi(y) / e^{\sqrt{2}y} = +\infty$$

whereas W' is p-integrable for any $p < 1$. This concludes the proof.

It is possible to prove the a.s. convergence of the martingale W' without passing through the martingales M_ϕ but this would require a stopping argument on the branching trees. □

In the next paragraph we shall attempt to better understand the limiting variable W'.

5. A PASSAGE PROBLEM FOR THE BROWNIAN TREE.

Let us consider the dyadic brownian tree (issued from 0) of the first paragraph and its "first crossing" of lines of equations $x = \lambda t - s$ in the x,t-plane $(\lambda, s \in R_+^*)$. We say that a particle u from the tree which is at the point $X_u(t)$ at time t provided it is then living $(T_u \le t < T_u + \tau_u)$, crosses the line $x = \lambda t - s$ for the first time if $X_u(t) = \lambda t - s$ and if for every $t' < t$ the ancestor v of u living at that time t' ($v = u$ may be) is such that $X_v(t') < \lambda t' - s$. The tree drawn below has three first crossings with the drawn line.

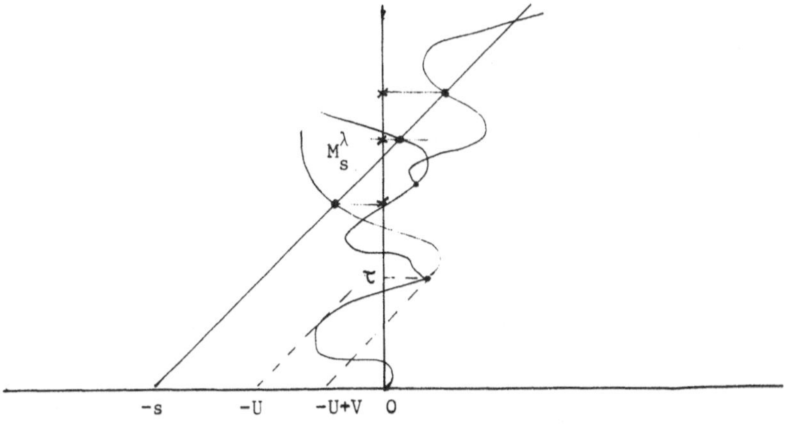

Fig. 1

The number Z_s^λ $(\lambda, s \in R_+^*)$ of these first crossings is a random variable with values in $\bar{N} = \{0,1,2,\dots,+\infty\}$. We shall restrict ourselves to the case where $\lambda \geq \sqrt{2}$ (since in the case $0 \leq \lambda < \sqrt{2}$ the following arguments would show that $Z_s^\lambda = +\infty$ a.s.); under this condition $0 < Z_s^\lambda < \infty$ a.s. since $\lim_{t\to\infty} (N_t^* - \sqrt{2}t) = -\infty$.

<u>Proposition 3</u>. <u>For each</u> $\lambda \geq \sqrt{2}$, <u>the integer-valued process</u> $(Z_s^\lambda, s > 0)$ <u>is a Galton-Watson process without extinction whose infinitesimal generating function</u> a <u>is given by</u>

$$a = \psi' \circ \psi^{-1} \quad \text{on} \quad]0,1[$$

<u>where</u> $\psi : R \to]0,1[$ <u>is the solution of Kolmogorov's equation</u>

$$\frac{1}{2}\psi'' - \lambda\psi' = \psi(1-\psi)$$

(ψ is uniquely determined up to a translation which of course does not modify $\psi' \circ \psi^{-1}$).

<u>Proof</u>. The branching property of the $Z_\bullet^\lambda$-process follows easily form the strong Markov property of the Brownian motion since each particle of the dyadic brownian branching tree starts afresh at its first crossing (if any)... A detailed formalism to prove rigorously this statement has been developped by B. Chauvin [ ] . The Galton-Watson process $Z_\bullet^\lambda$ presents no extinction and is certainly non degenerate; let us then proceed to determine the ψ function attached to it by paragraph 1.

We shall rely on the following lemma for the brownian motion which is due to D. Williams [14]; its second part will only be used later.

<u>Lemma</u>. <u>Given a real brownian motion</u> $(B(t), t \in R_+)$ <u>and an independent mean</u> 1 <u>exponential time</u> τ, <u>the two positive random variables</u>

$$(5.1)\qquad U = \max_{0\le t\le \tau} (\lambda t - \beta(t)) \ , \ V = U - (\lambda\tau - \beta(\tau))$$

<u>are independent and exponentially distributed with respective parameters</u> $\lambda_- = \sqrt{\lambda^2+2}-\lambda$ <u>and</u> $\lambda_+ = \sqrt{\lambda^2+2}+\lambda$.

<u>Moreover</u> $U = \lambda\tau_o - \beta(\tau_o)$ <u>for an unique</u> $\tau_o \in [0,\tau]$ <u>a.s. and the conditional law of</u> $(\tau_o, \tau-\tau_o)$ <u>with respect to</u> U,V <u>is given by</u>

$$E[\exp(-\alpha\tau_o - \beta(\tau-\tau_o))/U,V] = \exp[-U\rho(\alpha) - V\rho(\beta)]$$

<u>where</u> $\rho(\alpha) = (\lambda^2+2+2\alpha)^{1/2} - (\lambda^2+2)^{1/2}$ <u>and</u> $\alpha,\beta \in R_+$.

Consider again the Galton-Watson $z^\lambda_\cdot$ and let β and τ in the preceding lemma be the motion performed by the initial particle and its death time. Then clearly $z^\lambda_s = 1$ for $s < U$. Furthemore at its death time τ the initial particle lies V to the right of the line $x = \lambda t - U$, so that if $z'^\lambda_\cdot$ and $z''^\lambda_\cdot$ denote the $z^\lambda_\cdot$-process for the two descendants of the initial particle

$$(5.2)\qquad z^\lambda_U = z'^\lambda_V + z''^\lambda_V$$

In this formula V, $z'^\lambda_\cdot$ and $z''^\lambda_\cdot$ are independent and the processes $z'^\lambda_\cdot$ and $z''^\lambda_\cdot$ have the same law as $z^\lambda_\cdot$.

These results lead to an equation for the infinitesimal generating function and for the ψ-function attached to $z^\lambda_\cdot$ by (1.1). Indeed by the preceding lemma

$$\alpha = \lambda_- ,$$

$$\sum_1^\infty p_k u^k = E[u^{z^\lambda_U}] = E[u^{z'^\lambda_V} u^{z''^\lambda_V}]$$

$$= \int_0^\infty dv\ \lambda_+ e^{-\lambda_+ v} [E(u^{z^\lambda_v})]^2$$

Hence, since $\psi' = a \circ \psi = \alpha(\sum_1^\infty p_k \psi^k - \psi)$ we obtain that

$$(I + \frac{1}{\lambda_-} \frac{d}{du}) \psi(u) = \sum_1^\infty p_k \psi(u)^k$$

$$= \int_0^\infty dv\, \lambda_+ e^{-\lambda_+ v} \psi(u+v)^2$$

using 1.2, and this implies that

$$(5.3) \quad \frac{1}{2} \psi'' - \lambda\psi' - \psi = -(I - \frac{1}{\lambda_+} \frac{d}{du}) (I + \frac{1}{\lambda_-} \frac{d}{du}) \psi(u) = -\psi^2(u)$$

The proof of proposition 3, is then completed due to the uniqueness of the solution $\phi : R \to]0,1[$ of Kolmogorov's equation. □

By the results recalled in paragraph 1, $(\psi(u-s)^{Z_s^\lambda}, s \in R_+)$ is a martingale for each u with a limit of the form $\exp(-e^{au} W)$ where $a = \lambda - \sqrt{\lambda^2 - 2}$ (i.e. $v(a) = a\lambda$) and where the law of W is determined by $E[\exp(-e^{ua} W)] = \psi(u)$ $(u \in R)$. Furthemore for the most interesting case $\lambda = \sqrt{2}$ (where $a = \sqrt{2}$) the reproduction law of the Galton - Watson process $Z_\cdot^{\sqrt{2}}$ does not belong to the $L \log L$ class as the behavior of ψ at $-\infty$ shows and

$$(5.4) \qquad \text{a.s.} \lim_{s \to \infty} s\, e^{-\sqrt{2}s} Z_s^{\sqrt{2}} = W$$

where the term s has to be emphasized (up to a constant which can be set equal to 1 by a proper choice of ψ).

The preceding function ψ coïncides with the function ϕ of paragraph 4 and in fact so do the variables W and W' because the one-parameter martingales considered in the two paragraphs,

$\left(\prod_{x \in N_t} \phi(x-\lambda t)\,,\ t \in R_+\right)$ and $\left((\psi(-s))^{Z_s^\lambda}\,,\ s \in R_+\right)$ for example, may be obtained by stopping "tree martingales" when the particles of the branching tree either reach time t (paragraph 4) or reach the space - time line $x = \lambda t - s$. For the definition and study of "tree - martingales" see the forthcoming paper [8] by B. Chauvin.

Instead of only looking at the numbers Z_s^λ of first crossings of the line $x = \lambda t - s$ $(s \in R_+)$ by the brownian tree, let us consider the point process M_s^λ on R_+ consisting of the times t at which these first crossings occur (see Fig. 1). The right-continuous family $(M_s^\lambda\,,\ s \in R_+)$ of point processes on R_+ is then a spatial branching process, homogeneous in space (space for M_s^λ is time for the initial tree !) for the same reasons that their total masses $M_s^\lambda(R_+) = Z_s^\lambda$ build a Galton - Watson process. It does not seem possible to find explicitedly the (non linear) infinitesimal generator of M^λ but some of the results above concerning Z^λ are easily extended to M^λ as follows.

Let T_s^λ be the first time the initial particle which performs a stopped brownian motion $(\beta(t)\,,\ 0 \le t \le \tau)$ reaches the line $x = \lambda t - s$ $(s \in R_+)$; T_s^λ is only properly defined is $s < U$ and it is well known that

$$E[\exp(-\alpha\, T_s^\lambda) / s < U] = \exp[-s\, \rho^\lambda(\alpha)]$$

where ρ^λ denotes the Levy function

$$\rho^\lambda(\alpha) = (\lambda^2 + 2 + 2\alpha)^{1/2} - (\lambda^2 + 2)^{1/2} \qquad (\alpha \in R_+)$$

Then

$$(5.2') \qquad M_s^\lambda = \delta_{T_s^\lambda} \quad \text{for} \quad s < U$$

whereas

$$(5.2'') \qquad M_U^\lambda = \delta_\tau * (M_V'^\lambda + M_V''^\lambda)$$

where $M_\bullet'^\lambda$ and $M_\bullet''^\lambda$ are the $M_\bullet^\lambda$ processes attached to the two descendants of the initial particle; these two processes are independent of $(U,V,\tau,T_\bullet^\lambda)$ and follow the same law as M^λ.

These last formulas are intimately related to the decomposition of the parabolic operator

$$L = \frac{\partial}{\partial t} + \frac{1}{2}\frac{\partial^2}{\partial x^2} - \lambda\frac{\partial}{\partial x} - I$$

into

$$(5.3') \qquad L = \frac{1}{\lambda_- \lambda_+}\left(\frac{\partial}{\partial x} + \lambda_- + J^\lambda\right)\left(\frac{\partial}{\partial x} - \lambda_+ - J^\lambda\right)$$

where

$$J^\lambda = \left(\lambda^2 + 2 - 2\frac{\partial}{\partial t}\right)^{1/2} - (\lambda^2 + 2)^{1/2}$$

is in fact the infinitesimal generator of the T^λ-subordinator. Indeed for every μ the function $\Phi_\mu(x,t) = \Phi_\mu(x + (\lambda-\mu)t)$ is a solution of the equation

$$L\Phi_\mu + \Phi_\mu^2 = 0$$

since $\frac{1}{2}\Phi_\mu'' - \mu\Phi_\mu' - \Phi_\mu(1-\Phi_\mu) = 0$ and this last equation may be rewritten

$$\frac{1}{\lambda_-}\left(\frac{\partial}{\partial x} + \lambda_- + J^\lambda\right)\Phi_\mu(x,t) = \int_0^\infty \lambda_+ e^{-\lambda_+ v}\, dv\, E[\Phi_\mu^2(x+v\, ,\, t+T_v)/v < U]$$

by what precedes and by inverting $\frac{1}{\lambda_+}\left(\frac{\partial}{\partial x} - \lambda_+ - J^\lambda\right)$; but this equation expresses analytically the martingale property of the martingale $\left(\prod_{t \in M_s^\lambda} \Phi_\mu(x + (\lambda-\mu)t)\, ,\, s \in R_+\right)$ taking (5.2') and (5.2") into account.

BIBLIOGRAPHY.

[1] S. Asmussen and H. Hering. Branching Processes. Progress in Probability and Statistics 3. Birkhauser 1983.

[2] K. Athreya and P. Ney. Branching Processes. Springer 1972.

[3] J.D. Biggins. Martingale convergence in the Branching random Walk. Adv. Appl. Proba. 10 (1978) 62-84

[4] J.D. Biggins. Growth Rates in the Branching Random Walk. Z.f.W. 48 (1979) 17-34.

[5] M. Bramson. The convergence of solutions of the Kolmogorov nonlinear diffusion equations to travelling waves. Mem. Amer. Math. Soc. 44 (1983) n°285.

[6] B. Chauvin. Arbres et Processus de Bellman-Harris. Ann. I.H.P. 22 (1986).

[7] B. Chauvin and A. Rouault. The K.P.P. equations and branching brownian motion in the subcritical and critical areas. Application to spatial trees. To appear.

[8] B. Chauvin. Product martingales and stopping lines. To appear.

[9] R. Durrett and T.M. Liggett. Fixed Points of the Smoothing Transformation.

[10] N. Ikeda, M. Nagasawa and S. Watanabe. Branching Markov Processes. J. Math. Kyoto Univ. 8 (1968) 233-278 and 9 (1969) 95-160.

[11] A. Joffe. Oral communication.

[12] J. Neveu. Arbres et Processus de Galton - Watson. Ann. I.H.P. 22 (1986).

[13] K. Uchiyama. Spatial Growth of a branching process of particles living in R^d. Ann. of Proba. 10 (1982) 896-918.

[14] D. Williams. Path decomposition and continuity of local time for one-dimensional diffusions, I. Proc. London Math. Soc. (3) 28 (1974), 738-768.

ENERGY AND POTENTIALS

by

Zoran R. Pop-Stojanović

INTRODUCTION. This paper will show that under certain analytic condition concerning the potential operator U of a transient Hunt process X which is in duality with a Hunt process $\hat{X}$, the limit potential of a sequence of potentials which are bounded in energy and which belong to the class (D), will also belong to the class (D). All the assumptions and definitions concerning this duality are as in [1], Chapter 6. Other notations and definitions concerning the energy are as in papers [3],[4] and [5].

SETTING. Given two Hunt processes X and $\hat{X}$ which are in duality. Let $\zeta, \hat{\zeta}$, denote their life times.

Definition. An excessive function s which is finite a.e. is class (D) if whenever $T \uparrow T_n \geq \zeta$, as $n \to \infty$, T_n, T stopping times,

$$P_{T_n} s \downarrow 0 \quad \text{a.e.}$$

Proposition 1. If $s = \sum_1^\infty s_m$ and s_m are class (D) and $s < \infty$ a.e., then s is in class (D).

Proof. Let $T_k \uparrow T \geq \zeta$. Let $0 < g \in L^1$, $(s,g) = \int s(x)g(x)dx < \infty$ $\int g = 1$ and let $Q(.) = \int P^x(.)g(x)dx$.

Now

$$\infty > (s,g) = \sum_1^\infty (s_m, g) \quad .$$

Given $\varepsilon > 0$ choose m_ε such that:

$$\sum_{m_\varepsilon}^\infty (s_m, g) < \varepsilon \quad .$$

Then for all k,

$$\sum_{m_\varepsilon}^\infty (P_{T_k} s_m, g) < \varepsilon \quad .$$

In particular, one has

$$(P_{T_k} s, g) - \sum_1^{m_\varepsilon} (P_{T_k} s_m, g) < \varepsilon \ ,$$

uniformly in k. Letting $k\uparrow\infty$, and using that s_m are in class (D), we get

$$\lim_k (P_{T_k} s, g) < \varepsilon \quad ,$$

i.e., s is in class (D). □

Proposition 2 . Let $s\lessgtr\infty$ a.e. and $s = U\mu$. If μ does not charge polar sets then s is in class (D).

Proof . [Same as the proof of Theorem 3.4. p. 281 of [1]]. Indeed, let $0 < g\in L^1$ be such that $(s,g)<\infty$. Claim : the set of y's such that $\lim_n \int g(x)P_{T_n} u(x,y)dx \geq \varepsilon$ is polar. Indeed, if K is a compact subset of this set, $t = P_{T_k} 1 = U\nu$, then $P_{T_k} 1$ is in class (D). [Because $\lim_n E^x[P^{X_{T_n}}[T_k < \infty] : T_n < \zeta] = \lim_n P^x[T_k(\theta_{T_n})<\infty : T_n<\zeta]$ $= P^x[X_{\zeta-} \in K] = 0$] .

Therefore,

$$0 = \lim_n (g, P_{T_n} t) = \lim_n \int \nu(dy)\, P_{T_n} u(x,y)g(x)dx \geq \nu(1) \; .$$

An excessive function h will be called harmonic if

$$P_{T_n} h \downarrow 0 \quad \text{a.e.}$$

where $T_n = T_{K_n^c}$ and K_n are compact sets increasing to the state space. □

Proposition 3 . Suppose there is a sequence of potentials, bounded in energy and tending a.e. to a harmonic function. Then we can find a harmonic function h and a sequence Uf_n increasing to h such that sequences $||Uf_n||_e$ and $\int f_n$ are both bounded.

Proof . Suppose $p_n \to \tilde{h}$ a.e. and $\tilde{h}$ is harmonic and $p_n|| \leq M$. For any $g\geq 0$ such that $||Ug||_e \leq 1$ we have

$$(h,g) \leq \liminf_n (p_n,g) \leq M||Ug||_e \; .$$

Suppose $s_n = Uf_n \uparrow h$. Then, from above

$$(s_n, g) \leq M||Ug||_e$$

so that $||s_n|| \le 2M$, say. Let $h(x_o) = 2\varepsilon$ and

$$A = \{x: h>\varepsilon \}, \quad B = \{x: h \le \varepsilon \} .$$

By choosing subsequences if necessary, assume Uf_n1_A and Uf_n1_B converge a.e. to excessive functions h_1, h_2, respectively. Then, since $Uf_n1_B \le h \le \varepsilon$ on B and hence everywhere, we see that $h_2 \le \varepsilon$. Since $h_1 + h_2 = h$ and $h(x_o) = 2\varepsilon$ we see that $h \ne 0$. We have h_1 harmonic and

$$(f_n1_A, h) \le \lim_m (f_n1_A, Uf_m) \le 2M .$$

But $h \ge \varepsilon$ on A, so

$$\varepsilon \int f_n 1_A \le 2M .$$

Thus, h is the required function.

Let Ug_n increase to h_1 . Then $\int g_n$ increases and

$$\lim_n \int g_n \le \liminf_n \int f_n 1_A .$$

This completes the proof. □

Proposition 4 . Assume there is $\phi > 0$ such that $\hat{U}\phi$ tends to zero at infinity. Then, if p_n are potentials such that $||p_n||_e$ is bounded, $\lim_n p_n = h$ with h harmonic implies $h = 0$.

Proof . Using the previous Preposition we may assume

$$s_n = Uf_n \uparrow h , \ h \text{ harmonic and } ||s_n||_e \le M , \int f_n \le M.$$

Let ϕ be as in the hypothesis. We may assume that $\hat{U}\phi$ (and hence $U\phi$) has energy equal to 1. Let K be any compact set. Then $(h,\phi) = (P_{K^c}h,\phi) = \lim_n (P_{K^c}Uf_n,\phi) = \lim_n (f_n, \hat{P}_{K^c}\hat{U}\phi) \le ||\hat{P}_{K^c}\hat{U}\phi||_\infty M$, since $\int f_n \le M$. Since $\hat{U}\phi \to 0$ at infinity, for large K, $||\hat{P}_{K^c}\hat{U}\phi||_\infty$ is small. Thus $(h,\phi) = 0$ and hence $h = 0$. □

Theorem 5 . Let the assumptions be as in the previous Proposition. Assume p_n is a sequence of potentials which is bounded in energy. If p is excessive and $\lim_n p_n = p$ a.e., then p is in class (D).

Proof . By a result in [2] we can write p as a sum of a potential and a harmonic function h. This harmonic function is an increasing limit of potentials of the form Ug_k. Now $||Ug_k||_e \le \liminf_n ||p_n||_e$. Hence, the Proposition 3 applies and we conclude that h = 0. Thus p is a potential.

Write $p = U\mu$ and let us show that μ cannot charge polar sets. If it did it will charge a compact polar set K. Let D_n be a sequence of open relatively compact sets decreasing to K. Then $\hat{C}(D_n) = C(D_n)$ capacity of D_n decreases to zero. We may assume $\Sigma\, (\hat{C}(D_n))^{\frac{1}{2}} < \infty$. Let $\hat{s}_n = \hat{P}_{D_n} 1$. Then, $\hat{s}_n$ is class (D) potential. It is easy to show using Meyer's energy formula that $||\hat{s}_n||_e^2 \le 2\hat{C}(D_n)$. Let $\hat{s} = \Sigma\, \hat{s}_n$. Then $\hat{s}$ has finite energy and $\hat{s} = \infty$ on K. Therefore

$$(\mu, \hat{s}) = \infty.$$

We will now show that $(\mu, \hat{s}) < \infty$. Let $\hat{s} = \hat{U}\nu$, where ν is the Revuz measure of $\hat{s}$. Since $\hat{s}$ is a class (D) potential ν cannot charge polar sets. Hence, using Proposition 2, $U\nu$ is a class (D) potential. We claim it has finite energy. Indeed, if Ug has finite energy then

$$(U\nu, g) = (\nu, \hat{U}g) \le ||\hat{U}\nu||_e\, ||\hat{U}g||_e = ||\hat{U}\nu||_e\, ||Ug||_e .$$

Validity of this inequality for all $g > 0$ such that Ug has finite energy implies that $U\nu$ has finite energy. Hence

$$\begin{aligned}(\mu, \hat{s}) = (\mu, \hat{U}\nu) = (U\mu, \nu) &\le \liminf_n (p_n, \nu) \le \liminf_n (p_n, U\nu)_e \\ &\le \liminf_n ||p_n||_e\, ||U\nu||_e < \infty,\end{aligned}$$

which is a contradiction. Thus, μ cannot charge polar sets and so $p = U\mu$ is a class (D) potential. This completes the proof. □

Remark . Unfortunately, the assumption in Theorem 5 does not cover the case of a bounded domain in R^d, $d \ge 3$ whose boundary has irregular (for the complement) points. To cover this case we give another set of conditions.

Assume U maps L^2 into L^2 and U is normal, i.e. U and $\hat{U}$ commute .

Lemma 6 . Let $h_n = U\mu_n$ be bounded in energy. Suppose $h_n \to 0$ in measure as $n \to \infty$. Then $h_n \to 0$ in energy as $n \to \infty$.

Proof . Let $f \geq 0$ be in L^1 and $||Uf||_e < \infty$. We have to show

$$(Uf, h_n)_e \to 0 \text{ , as } n \to \infty \text{ .}$$

We have

$$(Uf, h_n)_e = (h_n, f) + (f, \hat{U}\mu_n) \text{ .} \tag{1}$$

By [2] , (h_n^2, f) is bounded. Since $h_n \to 0$ in measure,

$$(h_n, f) \to 0 \text{ , as } n \to \infty \text{ .}$$

Now let us show that

$$\hat{U}\mu_n \to 0 \text{ in measure as } n \to \infty \text{ .}$$

Now $U\mu_n \in L^2$, and so is $\hat{U}U\mu_n$. Since $U\mu_n \downarrow 0$, a.e. one has

$$\hat{U}U\mu_n \to 0\text{, as } n \to \infty \text{ , for each } x$$

such that

$$\int \hat{U}(x,y)(U\mu_1)(y) < \infty \text{ .}$$

But $\hat{U}U\mu_n = U\hat{U}\mu_n$. So, $\int u(x,y)\hat{U}\mu_n(y) \to 0$, $(\hat{U}\phi, \hat{U}\mu_n) \to 0$, or

$$\hat{U}\mu_n \to 0 \text{ in measure.}$$

Again since $\hat{U}\mu_n$ is in $L^2(fdx)$, we get from (1) that

$$(Uf, h_n)_e \to 0 \text{ ,}$$

or that $h_n \to 0$ weakly in energy. Therefore, a convex combination of h 's converges to zero. Given $\varepsilon > 0$, there is a convex combination

$$||\Sigma \alpha_i h_{n_i}||_e < \varepsilon \text{ .}$$

If $n \geq \max(n_i)$, then $h_n \leq \Sigma \alpha_i h_{n_i}$. So $||h_n||_e \leq 2\varepsilon$.

Thus $h_n \to 0$ in energy, as $n \to \infty$. This completes the proof. □

We have the immediate

Corollary 7 . Let $g > 0$ and $\hat{U}g$ be of finite energy. Then

$$\lim ||\hat{P}_{K^c}\hat{U}g||_e = 0$$

where the limit is over the compacts K increasing to the state space.

Now we can state the last result.

Theorem 8. Let U be normal and assume U maps L^2 into L^2. If p_n are class (D) potentials bounded in energy and $p = \lim_n p_n$ a.e. with p excessive, then p is class (D).

Proof. First, let us show that p is a potential. Let h be its harmonic part. Then if $g > 0$, and $Uf_n \uparrow h$,

$$(h,g) = (P_{K^c}h,g) = \lim (P_{K^c}Uf_n,g) = \lim (f_n,\hat{P}_{K^c}\hat{U}g)$$

$$\leq \lim ||\hat{U}f_n||_e ||\hat{P}_{K^c}\hat{U}g||_e \, .$$

The right-hand-side is small if K is large by Corollary 7. Thus $h = 0$ and hence p is a potential. That p is in class (D) is proved as in Theorem 5. □

ACKNOWLEDGMENT. The author wishes to express his profound gratitude to Professor Murali Rao for his valuable suggestions concerning this paper.

REFERENCES

[1] R.M. Blumenthal and R.K. Getoor, Markov Processes and Potential Theory, New York, Academic Press, 1968.

[2] K.L. Chung and M. Rao, A new setting for Potential Theory, Ann. Inst. Fourier 30, 1980, 167-198.

[3] Z.R. Pop-Stojanovic and Murali Rao, Some Results on Energy, Seminar on Stochastic Processes 1981, 135-150, Birkhauser, Boston 1981.

[4] Z.R. Pop-Stojanovic and Murali Rao, Remarks on Energy, Seminar on Stochastic Processes 1982, 229-235, Birkhauser, Boston, 1983.

[5] Z.R. Pop-Stojanovic, Convergence in Energy and the Sector Condition for Markov Processes, Seminar on Stochastic Processes 1984, 165-172, Birkhauser, Boston, 1986.

Z.R. Pop-Stojanović
Department of Mathematics, University of Florida, Gainesville, Florida 32611

BROWNIAN BITRANSFORMS

by

Thomas S. Salisbury *

§0. Introduction

Doob's h-transforms provide an analytic setting within which to study conditioned Brownian motion. They let one associate a transition function to each excessive h, giving a class of transition functions which include those for conditioned Brownian motion among them. Thus if one understands a particular function h, given analytically, one can often answer probabilistic questions about the corresponding conditioned process.

For example, if P^{x_0} is the law of Brownian motion (X_t) started at the fixed point x_0, and killed upon leaving the interval $(0,1)$, we may wish to study the law Q of X conditioned to hit a set B, and killed at the last time it does so. We do so in terms of

$$h(x) = P^x(X_t \in B \qquad \text{for some} \qquad t > 0)$$

(where x is now an arbitrary initial point). The relation is that if $t > 0$ and $\tau \in \mathcal{F}_t^+$, then

$$Q(\tau, \zeta > t) = P^{x_0}(\tau h(X_t)/h(x_0),\ \zeta > t).$$

We will be interested in a similar class of objects, among which can be found the laws of *pairs* of independent processes, appropriately conditioned and killed.

* Research supported in part by NSERC grant A8000 and NSF grant MCS 83-01072.

We'll call these objects *h-bitransforms*. Davis and Salisbury [4] used them to extract information about the law of a pair of n-dimensional Brownian motions, whose Wiener sausages were conditioned to intersect. This turned out to be a bit trickier for a pair of process, than for just a single one; partly because of the lack of a strong Markov property, but as well because h need not uniquely determine the law of an h-bitransform. While it gives some information, it does not give enough to characterize the law of the conditioned process in question.

For concreteness, we'll usually deal with two independent one-dimensional Brownian motions, each killed upon leaving $(0,1)$. Our focus will be the structure of the class of Brownian bitransforms. Since uniqueness fails, the question of existence assumes greater importance. We'll prove an existence theorem in §3, and will devote §2 to examples of interesting pairs of processes that can be expressed as bitransforms.

Part of this work was done during a visit to Stanford University. The author would like to thank Professor K. L. Chung for his kind hospitality.

§1. Definitions

Write Ω_1 for the set of continuous paths ω with values in the interval $(0,1)$, and lifetime $\zeta(\omega)$. For $t \geq \zeta(\omega)$ we take $\omega(t) = \Delta$. Usually X_s and Y_t are the coordinate functions on Ω. If we are working on $\Omega = \Omega_1 \times \Omega_1$ then $X_s(\omega_1, \omega_2) = \omega_1(s)$ and $Y_t(\omega_1, \omega_2) = \omega_2(t)$. In this case, $\mathcal{F}_{s,t} = \sigma\{X_{s'}, Y_{t'}; s' \leq s, t' \leq t\}$, and $\mathcal{F} = \mathcal{F}_{\infty,\infty}$. We write P^x for the law of Brownian motion started at x and killed upon leaving $(0,1)$. $P^{x,y}$ is the law of an independent pair of such process, started at x and y respectively. $G(x,y)$ is the Green function for Brownian motion on $(0,1)$, and ${}_yP^x$ is the law of an h-transform of P^x by $h(x') = G(x', y)$.

For $u = (s,t) \in \mathbf{R}^2_+$ and $u' = (s',t') \in \mathbf{R}^2_+$, we write $u \leq u'$ for $s \leq s'$, $t \leq t'$ and $u < u'$ for $s < s'$, $t < t'$. We'll let 0 be either the number zero, or the zero vector in $\mathbf{R}^2$. Borrowing a term from R.Dudley, we call a set $\zeta \subset \mathbf{R}^2_+$ a *lower layer* if

(a) if $u \le v$ and $v \in \zeta$ then $u \in \zeta$;

(b) if $u < v_n$, $v_n \to u$, and $v_n \notin \zeta$ for any n, then $u \notin \zeta$ as well.

A *bipath* is a function of the form

$$\omega(s,t) = \begin{cases} (\omega_1(s), \omega_2(t)), & \text{(s,t)} \in \zeta \\ (\Delta, \Delta), & \text{(s,t)} \notin \zeta \end{cases}$$

where $\omega_1, \omega_2 \in \Omega_1$, ζ is a lower layer, and

$$\zeta(\omega_1) = \inf\{s; (s,0) \notin \zeta\},\ \zeta(\omega_2) = \inf\{t; (0,t) \notin \zeta\} \tag{1.1}$$

(so that ω_1 is the first component of $\omega(\cdot, 0)$ and ω_2 is the second component of $\omega(0, \cdot)$). In this case, we call ζ the *lifeset* of ω. There are natural shift operators θ_u for $u \in \mathbf{R}^2_+$; $\theta_{s,t}(\omega, \omega') = (\theta_s\omega, \theta_t\omega')$.

Write Ω_0 for the set of bipaths. Clearly, we can regard Ω_0 as a Borel subset of the Polish space $D(E) \times D(E) \times LL$, where $D(E)$ is the Skorokhod space on $E = [0,1] \cup \{\Delta\}$, and LL is the set of lower layers, given the topology $d(\zeta_1, \zeta_2) = \rho(\zeta_1 \Delta \zeta_2)$, where ρ is a probability measure on $\mathbf{R}^2_+$ charging all open sets.

We'll let $Z_{s,t}(\omega) = \omega(s,t)$ be the coordinate functions on Ω_0, and will abuse our notation by also writing X_s for the first component of $Z_{s,0}$, Y_t for the second component of $Z_{0,t}$, and $\mathcal{F}_{s,t} = \sigma\{X_{s'}, Y_{t'};\ s' \le s, t' \le t\}$. For convenience, if ℓ is a lower layer, we'll write $\mathcal{F}_\ell = \vee\{\mathcal{F}_u; u \in \ell\}$. Likewise, we'll write $P^{x,y}$ for the law on Ω_0, under which X and Y are independent, with laws P^x and P^y, and the lifeset is the rectangle $\zeta = [0, \zeta(X)) \times [0, \zeta(Y))$.

A *biexcessive* function $h(x,y)$ is one that is excessive in each component separately. In our case, it will be biconcave, but since much of this works for other processes, we'll stick to the general term. For Brownian motion in $\mathbf{R}^n$, Avanissian [1] showed that h is necessarily lower semicontinuous, so that in fact it is also excessive as a function of the pair (x,y). A *bipotential* is a biexcessive function of the form $h(x,y) = \int G(x,x')G(y,y')\mu(dx',dy')$. A function is *biharmonic* if it is harmonic in each variable separately.

Given a biexcessive function $h(x, y)$, we call a law Q on Ω_0 an *h-bitransform* starting from $z = (x, y)$, if for each $u \geq 0$ and $\Lambda \in \mathcal{F}_u^+$ we have

$$(1.2) \qquad Q(\Lambda, u \in \zeta) = P^z(\Lambda h(Z_u)/h(z), u \in \zeta).$$

§2. Examples

Our first two classes of examples generalize constructions that are well known in the case of a single process. Since the arguments are also the same, they have been omitted.

(a) Conditioning. Let $A \in \mathcal{F}$, and define $L = \{u \in \mathbf{R}_+^2; Z \circ \theta_v \in A$ for some $v > u\}$. Let W be Z killed off L, and let Q be the law of W, conditioned on the event $\{0 \in L\}$. Then Q is an h-bitransform by $h(z) = P^z(Z \circ \theta_v \in A$ for some $v > 0)$.

If $A = \{Z_0 \in B\}$, this generalizes the usual construction for one process, in which X is killed at the last exit time from B. These were the bitransforms used in Davis and Salisbury [4], where B was taken to be a neighborhood of the diagonal.

If $B = B_x \times B_y$, then L is the rectangule $[0, L(B_x; X)) \times [0, L(B_y; Y))$, (where in general $L(C, \omega)$ denotes the last exit time of ω from C). The simplest way of obtaining a non-rectangular L is to choose $B = B^1 \cup B^2$, where $B^i = B_x^i \times B_y^i$. Then L may be a rectangle, or a region with two corners. For example, we may have the following picture:

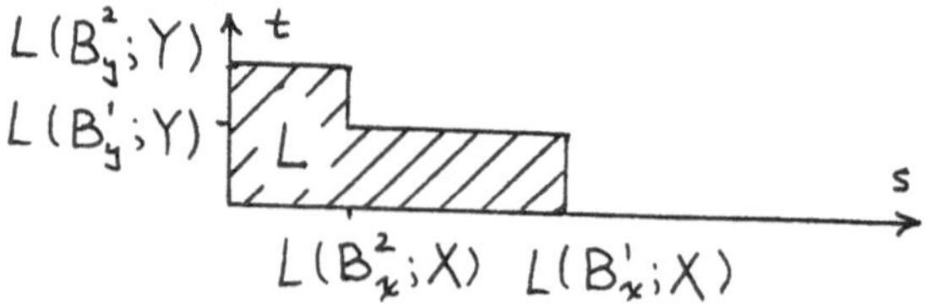

(b) Rectangles. The simplest lifesets are rectangular. If we restrict attention to bitransforms with rectangular lifesets, then a simple theory emerges, in complete analogy with the theory for a single process. It follows from a result of Cairoli [3], that the functions h that arise have a special form; they have an integral representation in terms of the product of the Martin kernels for X and Y. This class includes all bipotentials, and all biharmonic functions, but not all biexcessive functions. The corresponding h-bitransforms are uniquely determined by h (again, assuming rectangular lifesets). For example, in the case of a bipotential

$$h(x,y) = \int G(x,x')G(y,y')\mu(dx',dy')$$

we can build an h-bitransform as follows: starting from (x,y), choose a deathpoint $(X_{\zeta(X)-}, Y_{\zeta(Y)-})$ to have law $G(x,x')G(y,y')\mu(dx',dy')/h(x,y)$. Then let X and Y be conditionally independent, given deathpoint (x',y'), with conditional laws ${}_{x'}P^x$, ${}_{y'}P^y$. The general construction simply replaces Green kernels by Martin kernels. For a different construction in the case of a biharmonic h, see Walsh [8].

(c) Time Reversal. Part of the interest of bitransforms lies in the variety of ways the processes X and Y can be coupled together. In (b) they were independent except for a possible coupling via the deathpoint. In contrast, the following construction shows that one may in fact be a function of the other.

Fix x and y, and let X have law ${}_yP^x$. Let $Y = \hat{X}$, the reverse of X from its (finite) lifetime. Let $\zeta = \{(s,t); s+t < \zeta(X)\}$. The law of the resulting process is an h-bitransform, for $h(x,y) = G(x,y)$. The argument is simple; it (though not the statement of the result) may be found in Lemmas 3.1/3.3 of Davis and Salisbury [4].

(d) Skorokhod Embedding. The following example will give the non-uniqueness of h-bitransforms as a consequence of the fact that there may be many different schemes for embedding one measure in another.

The example is only half Brownian; we'll still take X to be Brownian motion on $(0,1)$, but we'll take Y to be a two state Markov process instead. Starting at a, it waits an exponential time, then jumps to b where it stays forever. Let the law of Y starting from a be R.

A biexcessive function h is now really a pair of excessive functions

$$f(x) = h(x,a), g(x) = h(x,b),$$

satisfying the condition

$$f \geq g. \tag{2.1}$$

Assume that f and g are potentials; $f = G\mu, g = G\nu$. As a consequence of (2.1), there is a stopping time T such that

$$P^{\mu}(X_T \in dx) = \nu(dx).$$

Let $L = \zeta(X) - T(\hat{X})$

$$X_t^L = \begin{cases} X_t, & t < L \\ \Delta, & t \geq L, \end{cases}$$

where again $\hat{X}$ is the reverse of X from its (finite) lifetime.

(2.2) LEMMA. *Under* ${}_fP^x$, X^L *is a* g*-transform, and*

$$g(x) = f(x)\ {}_fP^x(L > 0).$$

Proof. We'll only show the second statement, as the first then follows easily. By Proposition 1.3 of Walsh [7],

$$\begin{aligned} f(x)\ {}_fP^x(L>0) &= \int G(x,y)\ {}_yP^x(L>0)\mu(dy) \\ &= \int G(x,y)\ {}_xP^y(T<\zeta)\mu(dy) \\ &= \int G(x,y)P^y(G(X_T,x)/G(y,x), T<\zeta)\mu(dy) \\ &= P^{\mu}(G(X_T,x), T<\zeta) \\ &= \int G(z,x)P^{\mu}(X_T \in dz) = G\nu(x) = g(x). \end{aligned}$$

□

Now, under Q let X and Y be independent with laws ${}_fP^x$ and R. Let ζ be the set

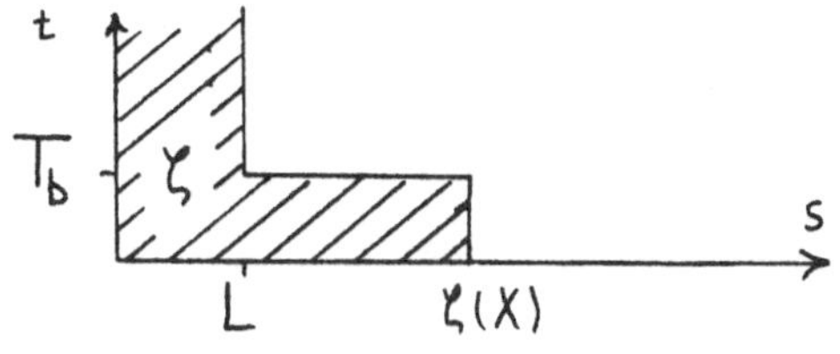

where T_b is the hitting time of b by Y. Then an immediate consequence of Lemma (2.2) is that Q is an h-bitransform of (P^x, R).

(e) Nonuniqueness. One can obtain an elementary example of nonuniqueness by randomizing the type of transform encountered in part (a). Let $0 < a < a' < 1$, $0 < b' < b < 1$. Let A_1 be $\{\text{hit}(a,b) \text{ or } (a',b')\}$ with probability $\frac{1}{2}$, and $\{\text{hit}(a,b) \text{ and } (a',b')\}$ with probability $\frac{1}{2}$. Let A_2 be $\{\text{hit}(a,b)\}$ with probability $\frac{1}{2}$, and $\{\text{hit}(a',b')\}$ with probability $\frac{1}{2}$. Performing the construction of part (a) with either A_1 or A_2 produces an h-bitransform by

$$\begin{aligned} h(x,y) &= \frac{1}{2}P^{x,y}(\text{hit}(a,b) \quad \text{or} \quad (a',b')) + \frac{1}{2}P^{x,y}(\text{hit}(a,b) \quad \text{and} \quad (a',b')) \\ &= \frac{1}{2}P^{x,y}(\text{hit}(a,b)) + \frac{1}{2}P^{x,y}(\text{hit}(a',b')), \end{aligned}$$

and these bitransforms are in fact different. For example, on the set where X and Y both die at 1, the lifesets are

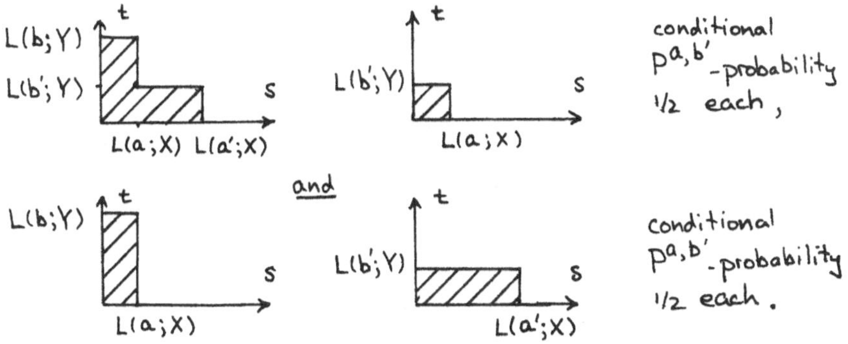

(f) Local Times. Let $a, b \in (0,1)$. We'll couple together ${}_aP^x$ and ${}_bP^y$, to give a bitransform by $G(x,a)G(y,b)$. Unlike the bitransform that the construction of (b) would give, X and Y will not be independent. Instead, we'll couple them so that $\ell_a(X) = \ell_b(Y)$, where ℓ_a and ℓ_b are local times at a and b respectively. Since the coupling shouldn't affect the marginal laws of $\ell_a(X)$ and $\ell_b(Y)$, we'll use a bizarre normalization of local time; namely, we'll scale local time so that regardless of $a \in (0,1)$, $\ell_a(X)$ has an exponential distribution of mean 1 under P^a. Write ℓ_a^s, ℓ_b^t for the corresponding local times, when the processes are observed just up to times s and t (rather than their lifetimes).

To carry out the coupling, let X have law ${}_aP^x$ under Q. Let $Q(Y' \in d\omega' \mid X = \omega) = {}_bP^y(Y' \in d\omega' \mid \ell_b(Y') \geq \ell_a(\omega))$. Let Y be Y' killed at the last time t that $\ell_b^t(Y') = \ell_a(X)$. Let

$$\zeta = \{(s,t);\ \ell_a^s(X) + \ell_b^t(Y) < \ell_a(X)\}.$$

LEMMA. *Q is a $G(\cdot,a)G(\cdot,b)$-bitransform.*

Proof. Let $\sigma \in \mathcal{F}_{s,0}^+$, $\tau \in \mathcal{F}_{0,t}^+$. Then

$$\begin{aligned}
Q(\sigma\tau,(s,t)\in\zeta) = {}& {}_aP^x(\sigma\ {}_bP^y(\tau, \ell_b^t(Y') + \ell_a^s(X) < \ell_a(X) < \ell_b(Y') \mid X) \\
& \quad / \ {}_bP^y(\ell_b(Y') > \ell_a(X) \mid X), s < \zeta(X)) \\
= {}& {}_aP^x(\sigma\ {}_bP^y(\tau e^{-(\ell_a(X)-\ell_b^t(Y'))}, \ell_a^s(X) < \ell_a(X) - \ell_b^t(Y') \mid X) \\
& \quad / e^{-\ell_a(X)}, s < \zeta(X)) \\
= {}& {}_bP^y(\tau e^{\ell_b^t(Y')}\ {}_aP^x(\sigma, \ell_b^t(Y') < \ell_a(X) - \ell_a^s(X) \mid Y'), \\
& \quad t < \zeta(Y')) \\
= {}& {}_bP^y(\tau e^{\ell_b^t(Y')}\ {}_aP^x(\sigma, e^{-\ell_b^t(Y')}, s < \zeta(X) \mid Y'), t < \zeta(Y')) \\
= {}& {}_bP^y(\tau, t < \zeta(Y))\ {}_aP^x(\sigma, s < \zeta(X)) \\
= {}& P^{x,y}(\sigma\tau G(X_s,a)G(Y_t,b)/G(x,a)G(y,b), (s,t) \in \zeta)
\end{aligned}$$

□

At first, this example seems unrelated to those of part (a). It is therefore perhaps curious that it can be written as a limit of such bitransforms. Indeed, let Q_k be the bitransform obtained by taking

$$A_k = \{\ell_a(X) + \ell_b(Y) \geq k\}.$$

In the local time scale, its lifeset is

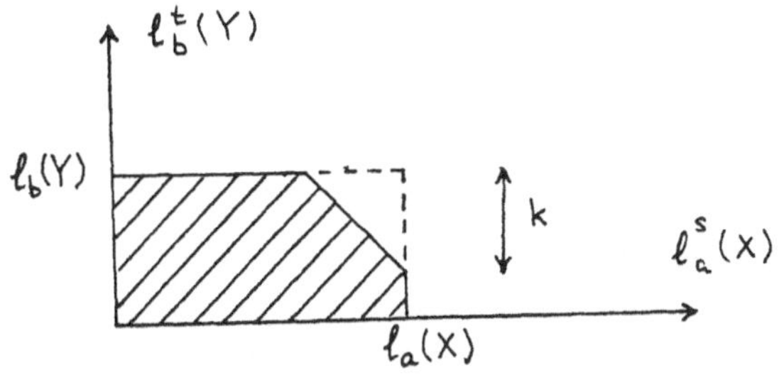

or, more typically (since $P^{a,b}(\ell_a(X) \geq k \mid \ell_a(X) + \ell_b(Y) \geq k) \to 0$ as $k \to \infty$)

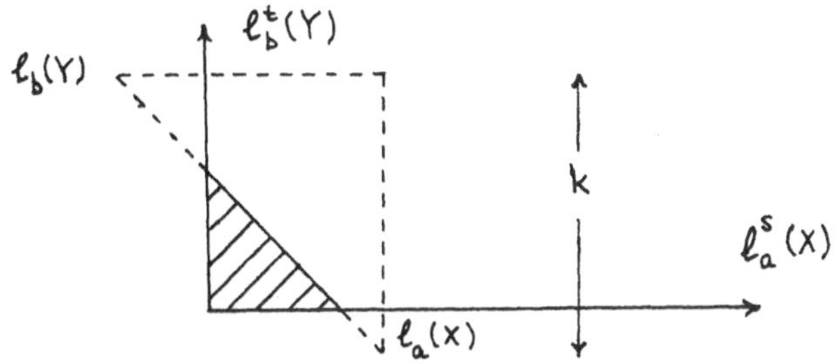

This makes it plausible that $Q_k \to Q$ in some sense. In fact (ignoring tightness), a computation similar to the one above shows that for $\sigma \in \mathcal{F}^+_{s,0}$, $\tau \in \mathcal{F}^+_{0,t}$

$$\begin{aligned} &Q(\sigma\tau, (s,0) \in \zeta, (0,t) \in \zeta) \\ &= P^{x,y}(\sigma\tau e^{\ell_a^s(X) \wedge \ell_b^t(Y)} G(X_s, a) G(Y_t, b)/G(x,a)G(y,b), (s,t) \in \zeta) \\ &= \lim_{k \to \infty} Q_k(\sigma\tau, (s,0) \in \zeta, (0,t) \in \zeta) \end{aligned}$$

which shows joint convergence of the finite dimensional distributions of X and Y under Q_k, to those under Q.

One can play the same game for other couplings. For example, with certain choices of a, a', b, b' one can couple so that the number of crossings of $[a, a']$ by X is dominated by the number of crossings of $[b, b']$ by Y. We'll spare the reader the details.

§3. Existence of Bitransforms

(3.1) THEOREM. *Let $h > 0$ be biexcessive, and let $x, y \in (0,1)$. then there is at least one bitransform starting from (x, y).*

For the remainder of the section, fix $h > 0$ and $x, y \in (0,1)$. All the Q's we write down will tacitly be assumed to satisfy $Q(X_0 \neq x \quad \text{or} \quad Y_0 \neq y) = 0$.

(3.2) LEMMA. *Let $(\mu, \mathcal{G})$ and $(\nu_n, \mathcal{H}_n)$ be finite measures, $n = 1, ..., m$. Assume that $\mathcal{G} \subset \cap \mathcal{H}_n$ and that $\mu \le \sum \nu_n$ on $\mathcal{G}$. Then there are measures $(\mu_n, \mathcal{H}_n)$ such that $\mu = \sum \mu_n$ on $\mathcal{G}$ and $\mu_n \le \nu_n$ on $\mathcal{H}_n$.*

Proof. Let $f \le 1$ be a Radon Nikodym density for $(\mu, \mathcal{G})$ with respect to $(\sum \nu_n, \mathcal{G})$. Set $\mu_n = f\nu_n$ on $\mathcal{H}_n$. □

We'll have to add a further layer of complexity to this result before we can use it:

(3.3) LEMMA. *Let $t' > t$, and let $s(n) \ge s$, $n = 1, ..., m$. Let μ be a finite measure on $\mathcal{F}_{s,t'}$. Let ν_n be finite measures on $\mathcal{F}_{s(n),t}$ such that $\mu \le \sum \nu_n$ on $\mathcal{F}_{s,t}$. Then there exist measures μ_n on $\mathcal{F}_{s(n),t'}$ such that $\mu = \sum \mu_n$ on $\mathcal{F}_{s,t'}$ and $\mu_n \le \nu_n$ on $\mathcal{F}_{s(n),t}$.*

Proof. In order to apply Lemma (3.2), we must first extend the ν_n to $\mathcal{F}_{s(n),t'}$. Let $\eta(\omega, d\omega')$ be a regular conditional probability for $(\mu, \mathcal{F}_{s,t'})$ given $\mathcal{F}_{s,t}$. That is: each $\eta(\omega, \cdot)$ is a subprobability on $\mathcal{F}_{s,t'}$; each $\eta(\cdot, A)$ is $\mathcal{F}_{s,t}$ measurable; for μ a.e. ω we have $\eta(\omega, A) = 1_A(\omega)$ for each $A \in \mathcal{F}_{s,t}$ (and for those ω for which this fails, $\eta(\omega, \cdot) \equiv 0$);

$$\mu(A \cap B) = \int_A \eta(\omega, B)\mu(d\omega) \qquad \text{for} \qquad A \in \mathcal{F}_{s,t}, B \in \mathcal{F}_{s,t'}.$$

Such an η will always exist, since $\mathcal{F}_{s,t'}$ is a separable Lusin measurable space (according to the definition of either Dellacherie and Meyer [5], or Blackwell [2]). In fact, for the μ we will apply the result to, an explicit form for η could be given.

Define ν_n' on $(\Omega \times \Omega, \mathcal{F}_{s(n),t} \otimes \mathcal{F}_{s,t'})$ by $\nu_n'(d\omega, d\omega') = \eta(\omega, d\omega')\nu_n(d\omega)$. Let $\varphi: \Omega \times \Omega \to \Omega$ be defined by $\varphi((\omega_1, \omega_2), (\omega_1', \omega_2')) = (\omega_1, \omega_2')$ and set $\nu_n'' = \varphi^{-1}\nu_n'$. If $A \in \mathcal{F}_{s(n),0}$ and $B \in \mathcal{F}_{0,t}$ then

$$\nu_n''(A \cap B) = \int_A \eta(\omega, B)\nu_n(d\omega) \le \int_A 1_B(\omega)\nu_n(d\omega) = \nu_n(A \cap B),$$

so that $\nu_n'' \le \nu_n$ on $\mathcal{F}_{s(n),t}$. If $A \in \mathcal{F}_{s,0}$ and $B \in \mathcal{F}_{0,t'}$ then

$$\begin{aligned}\mu(A \cap B) &= \int_A \eta(\omega, B)\mu(d\omega) \le \sum \int \eta(\omega, B)\nu_n(d\omega)\\ &= \sum \nu_n'(A \times B) = \sum \nu_n''(A \cap B).\end{aligned}$$

Thus $\mu \le \sum \nu_n''$ on $\mathcal{F}_{s,t'}$ so that Lemma (3.2) applies to μ and the ν_n'', showing the result. □

Now let $\mathbf{s}$ and $\mathbf{t}$ be finite subsets of $[0, \infty)$, each containing 0. Let $\Lambda(\mathbf{s}, \mathbf{t})$ consist of all bounded lower layers whose boundaries are contained in $(\mathbf{s} \times [0, \infty)) \cup ([0, \infty) \times \mathbf{t})$.

(3.4) LEMMA. *For* **s,t** *as above, there is at least one subprobability* Q *such that* $Q(\zeta \notin \Lambda(\mathbf{s}, \mathbf{t})) = 0$ *and (1.2) holds for* $u = (s, t)$, $s \in \mathbf{s}$ *and* $t \in \mathbf{t}$.

Proof. If $\mathbf{s} = \mathbf{t} = \{0\}$, let $\zeta \equiv \mathbf{R}_+^2$, $X \equiv x$, $Y \equiv y$ under Q. In general, we proceed by induction. Thus let Q work for given **s,t**. We'll construct Q' for $\mathbf{s}, \mathbf{t}'$ where $\mathbf{t}' = \mathbf{t} \cup \{t\}$, $t' > t$ for each $t \in \mathbf{t}$. Let $\mathbf{s} = \{s(0), ..., s(m)\}$ where $0 = s(0) < s(1) < ... < s(m)$, and write $s(m + 1) = \infty$. Write Λ for $\Lambda(\mathbf{s}, \mathbf{t})$ and Λ' for $\Lambda(\mathbf{s}, \mathbf{t}')$. Let $t^* = \sup\{t \in \mathbf{t}\}$. For $\ell' \in \Lambda'$, write $\lambda(\ell')$ for the unique element of Λ containing the same pairs (s, t) as ℓ', $s \in \mathbf{s}$ and $t \in \mathbf{t}$.

Let

$$\mathcal{G}_\ell = \vee\{\mathcal{F}_{s,t};\ s \in \mathbf{s}, t \in \mathbf{t}, (s, t) \in \ell\}$$

$$\mathcal{G}_{\ell'}' = \vee\{\mathcal{F}_{s,t};\ s \in \mathbf{s}, t \in \mathbf{t}', (s, t) \in \ell'\}.$$

Take $\nu_\ell(d\omega, d\omega')$ on $(\Omega, \mathcal{G}_\ell)$ to be $Q(\zeta = \ell, X \in d\omega, Y \in d\omega')$. We will construct measures $\mu_{\ell'}$ on $(\Omega, \mathcal{G}'_{\ell'})$ such that

$$\nu_\ell = \sum \{\mu_{\ell'}; \lambda(\ell') = \ell\} \qquad \text{on} \qquad \mathcal{G}_\ell, \tag{3.5}$$

and

$$\sum \{\mu_{\ell'}(d\omega, d\omega'); (s, t') \in \ell'\} = P^{x,y}(h(X_s, Y_{t'})/h(x, y), X \in d\omega, Y \in d\omega') \tag{3.6}$$
$$\text{on} \quad \mathcal{F}_{s,t'}, \quad \text{for} \quad s \in \mathbf{s}.$$

We can then set $Q'(\zeta = \ell', X \in d\omega, Y \in d\omega')$ to be $\mu_{\ell'}(d\omega, d\omega')$ on $\mathcal{G}'_{\ell'}$. For $u = (s, t)$, where $s \in \mathbf{s}$ and $t \in \mathbf{t}$, (1.2) then follows by induction and (3.5). For $u = (s, t')$ where $s \in \mathbf{s}$, (1.2) is exactly (3.6).

Note that strictly speaking we should define $Q'(\zeta = \ell', \cdot)$ on $\mathcal{F}_{\ell'}$, not just on $\mathcal{G}'_{\ell'}$. In fact, it can be extended to this set in an arbitrary manner, without disturbing the desired conditions. For example, we can take X to be constant on $[s(k), s(k+1))$ under Q', where $s(k) = \sup\{s \in \mathbf{s}; (s, 0) \in \zeta\}$ (and similarly for Y). Also note that, since Q' lives on Ω_0, we automatically have that $X \equiv \Delta$ on $[s(k+1), \infty)$ Q'-a.s. Though this will not be true $\mu_{\ell'}$-a.s. (the event is not even in $\mathcal{G}'_{\ell'}$), this again will not affect (1.2) for $s \in \mathbf{s}$ and $t \in \mathbf{t}'$, and so presents no obstruction to the construction of Q' (and likewise for Y).

Let $\pi_{s,t'}$ be the measure on $(\Omega, \mathcal{F}_{s,t'})$ given by the right hand side of (3.6). Let $\Lambda(i) = \{\ell \in \Lambda; (s(i), t^*) \in \ell, (s(i+1), t^*) \notin \ell\}$ (recall that $s(m+1) = \infty$). If $\ell \in \Lambda$, and $(s(i), t^*) \in \ell$, write

$$\ell(i) = (\ell \cap [0, \infty) \times [0, t')) \cup ([0, s(i+1) \times [0, \infty)) \in \Lambda'.$$

If $(0, t^*) \in \ell$, write $\ell-$ for $\ell \cap [0, \infty) \times [0, t')$. For $\ell \in \Lambda(j)$, the elements of Λ' having $\lambda(\ell') = \lambda$ are exactly $\ell-$ and $\ell(i)$, $i \le j$.

If $(0, t^*) \notin \ell$, we define $\mu_\ell = \nu_\ell$, giving (3.5) for ℓ of this form. The only element of $\Lambda(m)$ is $\ell = \mathbf{R}^2_+$, which has $\ell(m) = \mathbf{R}^2_+$ as well. Set

(3.7) $$\mu_{\ell(m)} = \pi_{s(m),t'} \quad \text{on} \quad \mathcal{F}_{s(m),t'} = \mathcal{G}'_{\ell(m)}.$$

Then automatically, $\mu_{\ell(m)} \leq \pi_{s(m),t^*} = \nu_\ell$ on $\mathcal{G}_\ell$.

Now $\pi_{s(m-1),t'} - \pi_{s(m),t'}$, is positive on $\mathcal{F}_{s(m-1),t'}$, and is dominated by

$$\begin{aligned}\pi_{s(m-1),t^*} - \pi_{s(m),t'} &= \sum \{\nu_\ell;\ (s(m-1),t^*) \in \ell\} - \pi_{s(m),t'} \\ &= \sum_{\Lambda(m-1)} \nu_\ell + \sum_{\Lambda(m)} (\nu_\ell - \mu_{\ell(m)}) \quad \text{on} \quad \mathcal{F}_{s(m-1),t^*}.\end{aligned}$$

Thus, by Lemma (3.3) we can find $\mu_{\ell(m-1)}$ on $\mathcal{G}'_{\ell(m-1)}$, for $\ell \in \Lambda(m-1) \cup \Lambda(m)$ such that

$$\begin{aligned}\pi_{s(m-1),t'} - \pi_{s(m),t'} = \sum_{\Lambda(m-1)\cup\Lambda(m)} \mu_{\ell(m-1)} \quad &\text{on} \quad \mathcal{F}_{s(m-1),t'}; \\ \mu_{\ell(m-1)} \leq \nu_\ell \quad \text{on} \quad \mathcal{G}_\ell \quad &\text{for} \quad \ell \in \Lambda(m-1); \\ \mu_{\ell(m-1)} \leq \nu_\ell - \mu_{\ell(m)} \quad \text{on} \quad \mathcal{G}_\ell \quad &\text{for} \quad \ell \in \Lambda(m).\end{aligned}$$

Iterate this procedure, to obtain $\mu_{\ell(i)}$ on $\mathcal{G}'_{\ell(i)}$, for $\ell \in \Lambda(j)$, $j \geq i$ such that

(3.8) $$\pi_{s(i),t'} - \pi_{s(i+1),t'} = \sum_{\Lambda(i)\cup\ldots\cup\Lambda(m)} \mu_{\ell(i)} \quad \text{on} \quad \mathcal{F}_{s(i),t'}, \quad \text{for} \quad i = 0, \ldots, m-1;$$

(3.9) $$\sum_{i=0}^{j} \mu_{\ell(i)} \leq \nu_\ell \quad \text{on} \quad \mathcal{G}_\ell \quad \text{for} \quad \ell \in \Lambda(j), j \leq m.$$

Then (3.6) follows from (3.7) and (3.8):

$$\begin{aligned}\pi_{s(i),t'} &= \pi_{s(m),t'} + \sum_{j=i}^{m-1} (\pi_{s(j),t'} - \pi_{s(j+1),t'}) \\ &= \sum \{\mu_{\ell(j)}; (s(j),t^*) \in \ell, j \geq i\} \\ &= \sum \{\mu_{\ell'}; (s(i),t') \in \ell'\} \quad \text{on} \quad \mathcal{F}_{s(i),t'}.\end{aligned}$$

By (3.9), we may set

$$\mu_{\ell-} = \nu_\ell - \sum_{i=0}^{j} \mu_{\ell(i)} \quad \text{on} \quad \mathcal{G}'_{\ell-} = \mathcal{G}_\ell, \quad \ell \in \Lambda(j).$$

This gives (3.5) for $\ell \in \Lambda$ with $(0,t^*) \in \ell$. □

Proof of Theorem (3.1). Recall that Ω_0 is a Borel subset of the Polish space $D(E) \times D(E) \times LL$, using which we will discuss weak convergence. As remarked above, the Q's of Lemma (3.4) may be taken to have X a.s. constant on $[s(k), s(k+1))$, where $s(k) = \sup\{s \in \mathbf{s};\ (s,0) \in \zeta\}$ (and similarly for Y). These measures are not bitransforms, but they are sufficiently close to being so that the argument of (7.8) of Davis and Salisbury [4] can be modified to show that they are tight. Extract a weakly convergent sequence Q_n along which $\mathbf{s}$ and $\mathbf{t}$ increase to $\mathbf{Q}_+$. Let Q be the limit.

Since Z is continuous on ζ Q_n-a.s. for each n, the same will be true Q-a.s. as well. A trivial estimate shows that for each $s, t > 0$

$$Q_n((s-\delta, t-\delta) \in \zeta, (s+\delta, t+\delta) \notin \zeta) \to 0 \quad \text{as} \quad \delta \downarrow 0, \quad \text{uniformly in} \quad n.$$

Thus $Q((s,t) \in \partial\zeta) = 0$ for each $s, t > 0$ and so, as for a single process, the finite dimensional distributions of $(Z_{s,t})_{s,t>0}$ under Q_n will converge to those under Q. We thus obtain (1.2) for $s, t \in \mathbf{Q} \cap (0, \infty)$. Another uniform approximation argument shows (1.2) for general $s, t \in [0, \infty)$.

One small point remains. Since Ω_0 is not closed, we still should show that Q lives on Ω_0. The problem is that $(\zeta(X), 0)$ or $(0, \zeta(Y))$ could conceivably not a.s. lie in the closure of ζ. Equivalently, the finite dimensional distributions of $(Z_{s,t})$ under Q_n might not converge to those under Q, when s or t are allowed to take on the value 0. This behavior is ruled out by yet another trivial uniformity: for each $s, t > 0$,

$$Q_n((s,0) \in \zeta, (s,\delta) \notin \zeta) \to 0 \quad \text{as} \quad \delta \downarrow 0, \quad \text{uniformly in} \quad n$$

$$Q_n((0,t) \in \zeta, (\delta,t) \notin \zeta) \to 0 \quad \text{as} \quad \delta \downarrow 0, \quad \text{uniformly in} \quad n$$

□

Remark. Though we have proven our theorem only for one dimensional Brownian motions, our argument will work more generally. Our use of weak convergence creates some artifical technical problems. For n-dimensional Brownian motion, these are easily handled, as in the appendix to Davis and Salisbury [4]. A more natural approrach would be to first choose a sequence of **s** and **t** increasing to $\mathbf{Q}_+$, to then select the Q_n to have convergence of finite dimensional distributions along this sequence, and then to use some analogue of Kolmogorov's theorem to obtain Q. We have avoided this, as the selection problems seemed much more delicate.

§4. Bibliography

[1] V. Avanissian. "Fonctions plurisousharmoniques et fonctions doublement sousharmoniques," Ann. Sci. Ec. Norm. Sup. 71 (1961), pp. 101-161.

[2] D. Blackwell. "On a class of probability spaces," Proc. 3rd Berkeley Symp. v.2 (1954), pp. 1–6.

[3] R. Cairoli. "Une représentation intégrale pour fonctions séparément excessives," Ann. Inst. Fourier, Grenoble 18 (1968) pp. 317–338.

[4] B. Davis and T. S. Salisbury. "Connecting Brownian paths," submitted to the Annals of Probability, 1987.

[5] C. Dellacherie and P. A. Meyer. Probabilités et Potentiel, Ch. I–IV, Hermann.

[6] J. L. Doob. Classical Potential Theory and its Probabilitistic Counterpart, Grund. der Math. Wiss. 262, Springer (1984).

[7] J. B. Walsh. "The cofine topology revisited," Proc. Symp. Pure Math. 31 (1977), pp. 131–151.

[8] J. B. Walsh. "Optional increasing paths," Lecture Notes in Mathematics 868 (1981), pp. 179–201.

Thomas S. Salisbury
Department of Mathematics
York University
North York, Ontario, CANADA

University of California, San Diego
La Jolla, California 92093, USA

ON TIME-REVERSAL OF REFLECTED BROWNIAN MOTIONS

by

R. J. WILLIAMS

§1. Introduction.

In this paper, a partial converse of a previous result on time-reversal of reflected Brownian motions is proved. Indeed, from [11], the time-reversal of a stationary reflected Brownian motion in a simple polyhedral domain is again a reflected Brownian motion if the directions of reflection satisfy a certain skew symmetry condition, and in this case the stationary distribution has a separable density. A converse of this result is proved here for a certain class of reflected Brownian motions in polyhedral domains that arise in applications to queueing theory. Before elaborating on this, some background on time-reversal of reflected Brownian motions is provided in the following paragraph.

A reflected Brownian motion (abbreviated as RBM) is a multi-dimensional diffusion process that behaves like Brownian motion with constant drift in the interior of a d-dimensional domain ($d \geq 1$) and is instantaneously reflected at the boundary where the direction of reflection is given by a non-tangential vector field on the boundary. When the domain is bounded and the boundary and reflection field are smooth, there is a unique stationary distribution for such an RBM, with a smooth density (relative to Lebesgue measure). Moreover, the time-reversal of the stationary RBM is also an RBM if and only if the stationary density is of the *exponential form* $p(x) = C\exp\{\eta \cdot x\}$, where $C > 0$ and η is a constant vector, both depending on the drift vector θ of the original process [8, 11]. The latter holds for all constant drifts θ if and only if the reflection field satisfies a certain skew symmetry condition [5]. In [11], a one-sided analogue of this was proved for reflected Brownian motions in simple polyhedral domains with a constant direction of reflection on each boundary face. These processes will be referred to as RBM's with polyhedral data. When the directions of reflection for such an RBM satisfy a skew symmetry condition, there is an invariant measure with an exponential form density and an associated dual process that is also an RBM with polyhedral data [11]. It follows that for a stationary RBM with polyhedral data satisfying the skew symmetry condition, the time-reversal of this process is also an RBM with polyhedral data.

In the sequel, a converse of the latter is proved for a class of RBM's with polyhedral data that arise as weak limits of *open* queueing networks under conditions of heavy traffic [9]. These processes will be referred to as *open* RBM's. It will be shown, with the aid of results in [5, 6, 11], that a stationary open RBM is also an open RBM under time-reversal if and only if the skew symmetry condition holds, and this is equivalent to the stationary distribution having a separable (or product form) density (cf. Theorem 1.1). It is reasonable to conjecture that an analogue of this result holds for other classes of RBM's with polyhedral data. One of the main obstacles to the proof of such a result is the lack of a general existence and uniqueness theorem for such processes.

For open RBM's, existence has been established by Harrison and Reiman [4]. The processes constructed in [4] were allowed to have correlated Brownian components, whereas those considered in the works quoted above have independent Brownian components. However, by performing a linear transformation of coordinates, one can see that the processes considered in [4] are equivalent to a certain class of RBM's with polyhedral data having independent Brownian components. For convenience, the representation of open RBM's given in [4] is used below for the precise statement and proof of the main result of this paper (Theorem 1.1).

The state space S of an open RBM is the d-dimensional positive orthant $\mathcal{R}_+^d$ and the data are specified by a triple (θ, Γ, R), where $\theta \in \mathcal{R}^d$ is a drift vector, Γ is a non-degenerate $d \times d$ covariance matrix (symmetric and positive definite) with ones on the diagonal, and R is a $d \times d$ reflection matrix of the form $R = I - P'$ where I is the $d \times d$ identity matrix and P' is the transpose of a $d \times d$ non-negative matrix P with zeros on the diagonal and having spectral radius less than one. Harrison and Reiman [4] have shown that given a d-dimensional Brownian motion $X = \{X(t), t \geq 0\}$, with drift vector θ, covariance matrix Γ, and satisfying $X(0) \in S$, there is a unique pair of d-dimensional processes (Y, Z) adapted to X such that

(i) $Z(t) = X(t) + RY(t) \in S \quad \text{for all} \quad t \geq 0,$

(ii) Y is a continuous non-decreasing d-dimensional process satisfying $Y(0) = 0$, and for each $k \in \{1, \ldots, d\}$, Y_k increases only at those times t for which $Z_k(t) = 0$, i.e., $\int_0^\infty 1_{\{Z_k(t) \neq 0\}} dY_k(t) = 0$.

Thus, Z is a semimartingale realization of the open RBM associated with the data (θ, Γ, R) and having the same initial distribution μ as X. The process Z behaves like a d-dimensional Brownian motion with drift vector θ and covariance matrix Γ in the interior of S. When the boundary face $\{x \in S : x_k = 0\}$ is hit, the kth component Y_k of Y increases causing an instantaneous displacement of the process Z in the direction given by the kth column of the reflection matrix R; the magnitude of the displacement is the minimal amount required to keep Z_k non-negative. Any process having the same law as Z will be referred to as an open RBM with data (θ, Γ, R) and initial distribution μ. For any $T > 0$, a process on $[0, T]$ is called an open RBM on $[0, T]$ if it is equal in law to an open RBM restricted to this interval. It is shown in §2 that there is a unique set of data (θ, Γ, R) associated with such an RBM. By allowing the initial position of Z to range over all states x in S, one obtains a continuous strong Markov process Z with an associated family of probability measures $\{P_x, x \in S\}$. This will be referred to as an open RBM with data (θ, Γ, R). Here the usual convention of using the same symbol to denote both a single process and a strong Markov family of processes has been adopted. This should cause no confusion since the meaning will be clear from the context.

Consider an open RBM Z with data (θ, Γ, R). It was shown in Harrison-Williams [6] that Z has a stationary distribution if and only if

$$\gamma \equiv -R^{-1}\theta > 0. \tag{1.1}$$

In this case, the stationary distribution π is unique and is absolutely continuous with respect to d-dimensional Lebesgue measure. It will be assumed henceforth that (1.1) holds and the density of π relative to Lebesgue measure will be denoted by p.

Definition. The stationary distribution for Z is of *product form* if and only if

$$p(x) = \prod_{k=1}^{d} p_k(x_k), \quad x = (x_1, \dots, x_d) \in \mathcal{R}_+^d, \tag{1.2}$$

where p_k is a probability density relative to Lebesgue measure on $\mathcal{R}_+$ for $k = 1, \dots, d$.

From [6], the stationary distribution for Z is of product form if and only if the following condition holds:

$$2\Gamma = 2I - P - P'. \tag{1.3}$$

Remark. This condition is the transformed version of a skew symmetry condition on the directions of reflection for an RBM with polyhedral data having independent Brownian components [5, 11]. It is slightly simpler than that given in [6], for the diagonal elements of Γ have been normalized to one here. This can always be achieved by performing a preliminary scaling in each coordinate direction.

For each fixed $T > 0$, define the reversed process Z_T^* on $[0, T]$ by

$$Z_T^*(t) = Z(T - t) \quad \text{for} \quad 0 \leq t \leq T.$$

Theorem 1.1. The following three conditions are equivalent.

(i) The stationary distribution for Z is of product form.

(ii) Condition (1.3) holds.

(iii) Suppose Z is initialized with its stationary distribution. Then for each $T > 0$, Z_T^* is an open RBM on $[0, T]$.

When these conditions hold, the stationary density p for Z has the form

$$(1.4) \qquad p(x) = C \exp\{-2\gamma \cdot x\}, \quad x \in S,$$

where γ is given by (1.1) and $C = 2^d \prod_{k=1}^d \gamma_k$. Furthermore, the data for the open RBM Z_T^* in (iii) is given by

$$((I - P)(I - P')^{-1}\theta, \ \Gamma, \ R' = I - P).$$

§2. Preliminaries.

In the following lemma, Z is a process with a fixed initial distribution.

Lemma 2.1. Suppose $T > 0$ and $\{Z(t), 0 \leq t \leq T\}$ is an open RBM on $[0, T]$. Then there is a unique set of data (θ, Γ, R) associated with this open RBM.

Proof. Without loss of generality, we may suppose that Z is the restriction to $[0, T]$ of an open RBM having a semimartingale representation of the form $X + RY$ as described in §1. It then suffices to show that R and the characteristics (θ, Γ) of X can be recovered from Z without a priori knowledge of X and Y. For this, we may assume that Z starts in the interior of S. For if not, in place of Z consider $\{Z(t+s), \ 0 \leq t \leq T - s\}$, which almost surely has this property for any $s \in (0, T)$ [6, Lemma 7.7].

Since (θ, Γ) are the infinitesimal characteristics of X, to determine them, it suffices to show that X can be recovered from Z. Since $\int_0^T 1_{\partial S}(Z(s))ds = 0$ a.s. [6, Lemma 7.2], by the L^2-isometry for stochastic integrals we have a.s. for all $t \in [0, T]$,

$$X(t) = \int_0^t 1_{S \setminus \partial S}(Z(s))dX(s).$$

But, since Y increases only on ∂S, the dX in the right member above may be replaced by dZ. It follows that X can be recovered (up to indistinguishability) from Z.

For the recovery of R, let $k \in \{1, \ldots, d\}$ and define

$$\tau_k = \inf\{t \geq 0 : Z_j(t) = 0 \quad \text{for some } j \neq k\}.$$

Then, since Y_j increases only when Z_j is zero, we have

$$Z(T \wedge \tau_k) - X(T \wedge \tau_k) = R_k Y_k(T \wedge \tau_k), \tag{2.1}$$

where R_k denotes the kth column of R. Now, Z starts in the interior of S (by assumption) and it behaves like X there, so there is a positive probability that Z reaches the kth face of S before reaching any other face and that it does so before time T. Since Y_k then almost surely increases, this yields $P(Y_k(T \wedge \tau_k) > 0) > 0$. Note that the left member of (2.1) is determined by Z. Thus, to prove R_k is determined by Z, it remains to show that $Y_k(T \wedge \tau_k)$ is so determined. This follows from setting $t = T$ and $\tau = \tau_k$ in (2.2) below.

Intuitively, Y_k is a measure of the local time of Z on the kth face of S. Indeed, it is proved below that for any stopping time τ for Z and $t \in [0, T]$,

$$Y_k(t \wedge \tau) = \lim_{\epsilon \downarrow 0} \frac{1}{2\epsilon} \int_0^{t\wedge\tau} 1_{[0,\epsilon]}(Z_k(s))ds, \tag{2.2}$$

where the limit is in L^2. For the proof, let $\epsilon > 0$ and define $g_\epsilon \in C^2(\mathcal{R}_+ \backslash \{\epsilon\}) \cap C^1(\mathcal{R}_+)$ such that $g_\epsilon''(u) = 1/\epsilon$ for $0 \leq u < \epsilon$, $g_\epsilon''(u) = 0$ for $u > \epsilon$, and for $u \in \mathcal{R}_+$, $g_\epsilon'(u) = \int_\epsilon^u g_\epsilon''(v)dv$ and $g_\epsilon(u) = \int_\epsilon^u g_\epsilon'(v)dv$. By applying Itô's formula to Z_k and suitable smooth approximations of g_ϵ, after passing to the limit of these approximations (cf. [1, p. 131]), one obtains a.s.

$$\begin{aligned} g_\epsilon(Z_k(t\wedge\tau)) - g_\epsilon(Z_k(0)) = & \int_0^{t\wedge\tau} g_\epsilon'(Z_k(s))dX_k(s) \\ & + \sum_{j=1}^d \int_0^{t\wedge\tau} g_\epsilon'(Z_k(s))R_{kj}dY_j(s) \\ & + \frac{1}{2}\int_0^{t\wedge\tau} g_\epsilon''(Z_k(s))ds. \end{aligned} \tag{2.3}$$

It follows from [10, Theorem 1] that a.s. $\int_0^T 1_{\{0\}}(Z_k(s))dY_j(s) = 0$ for all $j \neq k$, i.e., Y_j does not charge the intersection of face j with face k for $j \neq k$. Combining this with $(-g'_\epsilon) \downarrow 1_{\{0\}}$ as $\epsilon \downarrow 0$, it follows that the second last term in (2.3) tends a.s. to $-Y_k(t \wedge \tau)$ as $\epsilon \downarrow 0$. Furthermore, since $\int_0^T 1_{\{0\}}(Z_k(s))ds = 0$ a.s. [6, Lemma 7.2], by the L^2-isometry for stochastic integrals, the integral with respect to dX_k in (2.3) tends to zero in L^2 as $\epsilon \downarrow 0$. Finally, in this ϵ limit, $g_\epsilon \to 0$, and so (2.2) follows on letting $\epsilon \downarrow 0$ in (2.3). □

In the next lemma, some regularity properties of the stationary density p for an open RBM with data (θ, Γ, R) are proved. For this, let

$$L = \frac{1}{2}\sum_{i,j}\Gamma_{ij}\frac{\partial^2}{\partial x_i \partial x_j} + \sum_i \theta_i \frac{\partial}{\partial x_i},$$

and denote its formal adjoint by

$$L^\dagger = \frac{1}{2}\sum_{i,j}\Gamma_{ij}\frac{\partial^2}{\partial x_i \partial x_j} - \sum_i \theta_i \frac{\partial}{\partial x_i}.$$

The interior of the state space S will be denoted by G and the set of real-valued functions that are infinitely differentiable and have compact support in G will be denoted by $C_c^\infty(G)$.

Lemma 2.2. The density p is infinitely differentiable and strictly positive in G and satisfies

$$L^\dagger p = 0 \tag{2.4}$$

there.

Proof. It follows from the necessary condition developed in [6] that p satisfies the following relation

$$\int_G Lf(x)p(x)dx = 0 \quad \text{for all } f \in C_c^\infty(G),$$

where dx denotes integration with respect to Lebesgue measure on G. Viewing p as a Schwartz distribution on G, it follows from a version of Weyl's lemma for constant coefficient elliptic operators [7, Cor. 4.1.2], that p is infinitely differentiable in G and satisfies (2.4) there. Since p is non-negative and $L^\dagger$ is uniformly elliptic, it follows from (2.4) and the strong maximum principle [3, Theorem 3.5] that $p > 0$ in G. □

§3. Proof of Theorem 1.1.

Proof. Building on results established in [5, 11], it was shown in [6] that (i) and (ii) are equivalent. Indeed, using a Laplace transform relation for the stationary distribution, it was shown that (1.3) is necessary for the stationary distribution of an open RBM to be of product form. Conversely, after performing the appropriate linear transformation of coordinates to remove the correlation between the Brownian components, the sufficiency of (1.3) followed from [11], and the stationary density was shown to be given by (1.4). The proof in [11] proceeded via a family of approximating RBM's with smooth boundary data satisfying a skew symmetry condition. It would be useful to have a direct proof of the sufficiency of (1.3) that obviated this approximation procedure.

To close the circle of implications, it will be shown that (ii) implies (iii) and (iii) implies (i). To prove the former it suffices to show that, when (1.3) holds, there is an open RBM in duality with Z relative to π. For if Z and such a dual process are initialized with the stationary distribution π, then for each $T > 0$, the reversed process Z_T^* and the dual process are equivalent in law on $[0, T]$ (see for example [2]). The existence of such a dual process was established in [11], under the assumption of a skew symmetry condition, for RBM's with polyhedral data having uncorrelated Brownian components. To apply these results to the present situation, a linear transformation of coordinates must be performed to remove the correlation between the Brownian components of Z. The essentials of this translation procedure are given below; for further details see [6, 11].

Since Γ is real symmetric and positive definite, it has a decomposition of the form $\Gamma = U'AU$ where the rows of U are orthonormal eigenvectors of Γ, $U' = U^{-1}$

and A is the corresponding diagonal matrix of eigenvalues. Let $V = A^{-1/2}U$ and define $\tilde{Z} = VZ$, $\tilde{R} = VR$, $\tilde{N} = V^{-1}$, $\tilde{Q} = (I-P)V' - V^{-1}$ and $\tilde{\pi}(\,\cdot\,) = \pi(V^{-1}\,\cdot\,)$. Then $\tilde{Z}$ is an RBM with state space the polyhedral cone $\tilde{S} = \{\tilde{x} \in \mathcal{R}^d : \tilde{N}\tilde{x} \geq 0\}$ and data $(V\theta, I, \tilde{R})$. Thus, $\tilde{Z}$ behaves like Brownian motion with drift $V\theta$ and covariance matrix I in the interior of $\tilde{S}$, $\tilde{R} = \tilde{N}' + \tilde{Q}'$ where the rows of $\tilde{N}$ give the inward unit normals to the faces of $\tilde{S}$ and the rows of $\tilde{Q}$ give the corresponding tangential components of the directions of reflection for $\tilde{Z}$ on those faces. The measure $\tilde{\pi}$ defines a stationary distribution for $\tilde{Z}$ and condition (1.3) is equivalent to the skew symmetry condition

$$\tilde{N}\tilde{Q}' + \tilde{Q}\tilde{N}' = 0. \tag{3.1}$$

It was shown in [11] that when (3.1) holds, $\tilde{Z}$ has a dual process relative to $\tilde{\pi}$ that is an RBM in $\tilde{S}$ with data $(2(I - \tilde{N}^{-1}\tilde{Q})^{-1}V\theta - V\theta,\ I,\ \tilde{N}' - \tilde{Q}')$. Using the linear transformation V^{-1} to transform this dual process back to the orthant, one obtains an *open* RBM on S with data $(\hat{\theta}, \Gamma, \hat{R})$, where after algebraic simplification using $V'V = \Gamma^{-1}$,

$$\hat{\theta} = (I - P)(2\Gamma - (I - P))^{-1}\theta \quad \text{and} \quad \hat{R} = 2\Gamma - (I - P'). \tag{3.2}$$

Upon using (1.3) this reduces to

$$\hat{\theta} = (I - P)(I - P')^{-1}\theta \quad \text{and} \quad \hat{R} = I - P. \tag{3.3}$$

Notice that $\hat{\theta} = \hat{R}R^{-1}\theta$ and $\hat{R} = R'$. Thus, when (1.3) holds, the open RBM with data $(\hat{\theta}, \Gamma, \hat{R})$ is in duality with Z relative to π. As indicated above, this suffices to prove (ii) implies (iii).

Now suppose (iii) holds. For each $T > 0$, by Lemma 2.1, the data $(\theta^*, \Gamma^*, R^*)$ for the open RBM Z_T^* is uniquely determined, and by consistency it is the same for all $T > 0$. Let Z^* denote an open RBM with this data. Then, when Z and Z^* are initialized with distribution π, $Z^*|_{[0,T]}$ is equivalent in law to Z_T^*. It follows that for all continuous functions h and k having compact support in $\mathcal{R}^d$,

$$\int_S k(x)E_x[h(Z(t))]p(x)dx = \int_S h(x)E_x[k(Z^*(t))]p(x)dx \quad \text{for all } t \geq 0. \tag{3.4}$$

Here E_x denotes expectation when the initial state is x.

Let L and $L^\dagger$ be defined as in §2 and let

$$L^* = \frac{1}{2}\sum_{i,j}\Gamma^*_{ij}\frac{\partial^2}{\partial x_i \partial x_j} + \sum_i \theta^*_i \frac{\partial}{\partial x_i}.$$

By applying Itô's formula to a semimartingale representation of Z and a twice continuously differentiable function f with compact support in the interior G of S, one can deduce that for each $\lambda > 0$ and $h \equiv \lambda f - Lf$,

$$f(x) = E_x\left[\int_0^\infty e^{-\lambda t} h(Z(t))\,dt\right]. \tag{3.5}$$

Similarly, for a function g satisfying the same conditions as f and $k \equiv \lambda g - L^* g$,

$$g(x) = E_x\left[\int_0^\infty e^{-\lambda t} k(Z^*(t))\,dt\right]. \tag{3.6}$$

Taking Laplace transforms in (3.4) and using (3.5)-(3.6) yields after cancellation of like terms

$$\int_G L^* g\; pf\; dx = \int_G Lf\; pg\; dx.$$

After integrating by parts on the right and using the facts that f and g have compact support in G, and p is infinitely differentiable in G and satisfies $L^\dagger p = 0$ there (see Lemma 2.2), one obtains

$$\int_G L^* g\; pf\; dx = \int_G \left(L^\dagger g + \sum_{i,j}\Gamma_{ij} p^{-1}\frac{\partial p}{\partial x_i}\frac{\partial g}{\partial x_j}\right) pf dx.$$

Since $p > 0$ in G, by Lemma 2.2, and f and g are arbitrary C^2 functions with compact support in G, it follows that

$$L^* = L^\dagger + \sum_{i,j}\Gamma_{ij} p^{-1}\frac{\partial p}{\partial x_i}\frac{\partial}{\partial x_j}.$$

Equating coefficients yields,

$$\frac{\nabla p}{p} = \Gamma^{-1}(\theta^* + \theta)$$

and so $p(x) = \exp\{x \cdot \Gamma^{-1}(\theta^* + \theta)\}$. Hence, π is of product form. □

Acknowledgements. The author would like to thank J. M. Harrison for stimulating conversations on this and related work. The research reported here was supported in part by NSF Grant DMS 8319562-A01.

References.

1. Chung, K. L., and R. J. Williams, *Introduction to Stochastic Integration,* Birkhäuser, Boston, 1983.

2. Fitzsimmons, P. J., Homogeneous random measures and a weak order for the excessive measures of a Markov process, to appear in *Trans. Amer. Math. Soc.*

3. Gilbarg, D., and N. S. Trudinger, *Elliptic Partial Differential Equations,* Springer, New York, 2nd edition, 1983.

4. Harrison, J. M., and M. I. Reiman, Reflected Brownian motion on an orthant, *Ann. Prob.* **9** (1981), 302-308.

5. Harrison, J. M., and R. J. Williams, Multidimensional reflected Brownian motions having exponential stationary distributions, *Ann. Prob.* **15** (1987), 115-137.

6. Harrison, J. M., and R. J. Williams, Brownian models of open queueing networks with homogeneous customer populations, to appear in *Stochastics.*

7. Hormander, L., *Linear Partial Differential Operators,* Springer, New York, 1963.

8. Nagasawa, M., The adjoint process of a diffusion with reflecting barrier, *Kodai Math. Sem. Rep.* **13** (1961), 235-248.

9. Reiman, M. I., Open queueing networks in heavy traffic, *Math. Oper. Res.* **9** (1984), 441-458.

10. Reiman, M. I., and R. J. Williams, A boundary property of semimartingale reflecting Brownian motions, submitted.

11. Williams, R. J., Reflected Brownian motion with skew symmetric data in a polyhedral domain, to appear in *Probability Theory and Related Fields.*

R. J. WILLIAMS

Department of Mathematics

University of California at San Diego

La Jolla, CA 92093, U.S.A.

REMARKS ON HARMONIC FUNCTIONS AND INVARIANT MEASURES OF MARKOV PROCESSES

by

R. WU and M. LIAO

SUMMARY. Assuming duality and certain analytic conditions on the potential density $u(x,y)$, we show that any harmonic function is a constant and the invariant measure, if one exists, is unique.

We will assume that X_t is a standard process in duality with another standard process X'_t relative to some fixed reference measure m on the state space E. The reader is refered to [1, VI, Sec 1] for the precise definition of duality and to either [1] or [2] for other usual definitions. We will also assume that both X and X' are transient; i.e. for any compact K, $T_K \circ \theta_t \to \infty$ as $t \to \infty$ a.s.. However, only the transience of X is used in Proposition 1 and only that of X' is used in the rest of the discussion.

For simplicity, any subset A of E and any function f defined on E is automatically assumed to be measurable with respect to the Borel field of E.

There is a common potential density $u(x,y)$ which is excessive in x and co-excessive in y and satisfies:

$$Uf(x) = \int u(x,y)f(y)m(dy) \quad \text{and} \quad U'f(y) = \int m(dx)f(x)u(x,y)$$

for any $f \geq 0$, where U and U' are, repectively, the potential operators of X and X'.

A function $h \geq 0$ on E is said to be harmonic if for any compact subset K of E, $P_{K^c}h = h$.

For Brownian motion in R^n with $n \geq 3$, any harmonic function is a constant. Our first proposition generalizes this classical result under a general setting, which includes as special cases, not only Brownian motion, but also transient symmetric stable processes and one sided stable processes. Our proof which uses the general theory of Markov processes is quite short, compared with the complicated analytic proof for symmetric stable processes given in [3].

PROPOSITION 1. Assume: for any $x_1, x_2 \in E$ and $\varepsilon > 0$, there is a compact set K such that $u(x_1,y) \leq (1+\varepsilon)u(x_2,y)$ for $y \in K^c$. Then any harmonic function is a constant.

PROOF. Fix x_1, x_2 in E, $\varepsilon > 0$ and choose K as above. Let D_n be relatively compact open sets so that $D_n \uparrow K^c$. Let h be harmonic; then

$$h(x_1) = P_{K^c}h(x_1) = \lim_{j\to\infty} P_{K^c}(h\wedge j)(x_1)$$

$$= \lim_j \lim_n P_{D_n}(h\wedge j)(x_1).$$

Since X is transient, $P_{D_n}(h\wedge j)$ is a natural potential. By [1, (V)], it is the potential of a natural additive functional. By [6], we have

$$P_{D_n}(h\wedge j)(x) = \int u(x,y)v_n(dy),$$

where v_n is a measure supported by $\overline{D}_n$.

$$h(x_1) = \lim_j \lim_n \int u(x_1,y)v_n(dy)$$
$$\leq (1+\varepsilon)\lim_j \lim_n \int u(x_2,y)v_n(dy)$$
$$= (1+\varepsilon)P_{K^c}h(x_2) = (1+\varepsilon)h(x_2).$$

Since x_1, x_2 and $\varepsilon > 0$ are arbitrary, h has to be a constant. Q.E.D.

REMARK: If $u(x,y) > 0$, then the hypothesis of Proposition 1 means: $\lim_{y\to\Delta} u(x_1,y)/u(x_2,y) = 1$, where Δ is the cemetery point.

For any measure v on E, define

$$U'v(y) = \int v(dx)u(x,y).$$

LEMMA. If $U'v < \infty$ m-a.e., then $\lim_{t\to\infty} P'_t U'v = 0$ m-a.e..

PROOF. The duality implies:

$$\int P_t(x,dz)u(z,y) = \int u(x,z)P'_t(y,dz),$$

where P_t and P'_t are, respectively, the transition semigroups of X and X'. The above equality will be abbreviated to

$$P_t u(x,y) = uP'_t(x,y).$$

Since $U'v < \infty$ m-a.e., there is $g > 0$ such that $(g,U'v) = \int g(y)U'v(y)m(dy) < \infty$. We have

$$
\begin{aligned}
(g, P_t' U' v) &= \iint u P_t'(x,y) v(dx) g(y) m(dy) \\
&= \iint P_t u(x,y) v(dx) g(y) m(dy) \\
&= \int v(dx) P_t U g(x) = \int v(dx) E^x[\textstyle\int_t^\infty g(X_s) ds] \\
&\leq \int v(dx) E^x[\textstyle\int_0^\infty g(X_s) ds] = (g, U'v) < \infty.
\end{aligned}
$$

This implies: $\lim_{t\to\infty} P_t' U' v = 0$ m-a.e. Q.E.D.

A σ-finite measure v on E is said to be an invariant measure of X if $vP_t = v$ for any $t > 0$, where

$$vP_t(A) = \int v(dx) P_t(x,A).$$

The proof of (1.11) of [1, VI] shows that v is an invariant measure if and only if $dv = fdm$ and f satisfies $P_t' f = f$.

We need the Riesz decomposition of f to show that f is co-harmonic, i.e. harmonic with respect to X'. The duality assumption alone does not give us the desired decomposition. For this purpose, we may either assume, as in [1], that U' and $P_t' f$ are bounded continuous for any bounded and compactly supported $f \geq 0$, or, if we prefer to use only conditions on $u(x,y)$, assume the following conditions and apply the results of [5].

(1) m is a diffuse measure, i.e. m does not charge points.

(2) $(x,y) \to u(x,y)$ is lower semi-continuous and for each fixed y, $x \to u(x,y)$ is continuous on $E-\{y\}$.

Under the above two conditions, we can check that the hypotheses (i)-(iv) of [5] hold. By using the results of either [1, VI, Sec 2] or [5], we obtain

$$f = h + U'w$$

where h is co-harmonic and w is a Radon measure on E. Since $P'_t f = f$, so $P'_t U'w = U'w$. The above lemma implies $U'w = 0$. We have the following proposition.

PROPOSITION 2. Any invariant measure of X has a co-harmonic density with respect to m under one of the following assumptions:

(a) U' and P'_t, for any $t > 0$, have strong Feller properties.

(b) (1) and (2) hold.

Applying Proposition 1 to co-harmonic functions, we obtain the uniqueness of the invariant measure, which essentially contains the result of [4].

COROLLARY. Assume the hypotheses of Proposition 2 and that for any y_1, y_2 in E and $\varepsilon > 0$, there is a compact K such that $u(x,y_1) < (1+\varepsilon)u(x,y_2)$ for x in K^c. If v is an invariant measure of X, Then $v = cm$ for some constant c.

REMARK. By duality, m is an invariant measure of X if and only if X' is conservative, i.e. $P'_t 1 = 1$ for

$t > 0$. The following observation may be interesting: If 1 is co-harmonic, then m is an invariant measure of X. It is enough to show $P'_t 1 = 1$. By Theorem 4 of [2, 3.6], $1 = f_1 + f_2$, where $P'_t f_1 = f_1$ and $\lim_{t\to\infty} P'_t f_2 = 0$. f_2 is a natural potential, so $f_2 = U'w$ for some measure w on E. Since both 1 and f_1 are co-harmonic, the uniqueness of Riesz decomposition implies $f_2 = 0$.

REFERENCES

1. R. M. Blumenthal and R. K. Getoor, "Markov processes and potential theory", Academic Press, New York 1968.

2. K. L. Chung, "Lectures from Markov processes to Brownian motion", Springer-Verlag, Berlin 1982.

3. N. S. Landkof, "Foundation of modern potential theory", Springer-Verlag, 1972.

4. M. Liao and R. Wu, "The uniqueness of invariant measures of spatially homogeneous processes:, to appear in Chinese Acta Math.

5. M. Liao, "Riesz representation and duality of Markov processes", Lecture Notes in Math. 1123, 366-396.

6. D. Revuz, "Mesures associees aux fontionnelles additives de Markov", AMS Trans. 148, 1970, 501-531.

Department of Mathematics
Nankai University
Tianjin, P. R. China

Green Functions and Conditioned Gauge Theorem for a 2-Dimensional Domain

Z. Zhao

In this paper we investigate the properties of the classical Green function $G_D(\cdot,\cdot)$, i.e., the kernel function for the operator $(-\frac{\Delta}{2})^{-1}$, in a domain $D \subset R^2$, where D is a Jordan domain, namely, a bounded domain in R^2 with the boundary ∂D which consists of finitely many disjoint Jordan curves. It is easy to see that any bounded Lipschitz domain in R^2 is a Jordan domain.

Throughout the paper, D is supposed to be a Jordan domain and $G_D(\cdot,\cdot)$ is its Green function. We claim the main results as follows.

__Theorem 1__. There exists a constant $C = C(D) > 0$, such that for all $x, y \in D$,

$$\frac{1}{C}\,\ell n\left(1 + \frac{\rho(x)\rho(y)}{|x-y|^2}\right) \leq G_D(x,y) \leq C\,\ell n\left(1 + \frac{\rho(x)\rho(y)}{|x-y|^2}\right), \qquad (1)$$

where $\rho(x) = \mathrm{dist}(x,\partial D)$.

__Theorem 2__. (3G Theorem for d = 2). There exists a constant $C = C(D) > 0$ such that for all $x, y, z \in D$,

$$\frac{G_D(x,y)G_D(y,z)}{G_D(x,z)} \leq C[F(x,y)+F(y,z)], \qquad (2)$$

where $F(x,y) = \max(\ell n \frac{1}{|x-y|}, 1)$.

As what we did in [2] for $d \geq 3$ (joint paper with Cranston and Fabes), we can use Theorem 2 to prove the following conditioned Gauge Theorem:

Theorem 3. Let $q \in K_2^{loc}$. If the conditioned gauge

$$u(x,y) \equiv E_y^x[\exp\int_0^{\tau_D} q(X_s)ds] \not\equiv \infty,$$

in $D\times\overline{D}$ then there exists a constant $C = C(D,q) > 0$ such that

$$\frac{1}{C} \leq u(x,y) \leq C$$

for all $(x,y) \in D\times\overline{D}$.

We can also prove the boundary Harnack principle for the positive solutions of Schrödinger equation $(\frac{\Delta}{2} + q)U = 0$ by the exactly same arguments as in [2]. We don't want to repeat it here.

To prove Theorem 1 and 2 we need some preparatory properties. Let

$$B = (x \in R^2: |x| < 1),$$

$$B^* = (x \in R^2: |x| > 1)$$

and

$$S = (x \in R^2: |x| = 1).$$

For any Jordan curve Γ, we use $\mathrm{Int}(\Gamma)$ and $\mathrm{Ext}(\Gamma)$ to denote the interior and exterior region of Γ, respectively.

For two domains D_1 and D_2, ϕ is said to be an extended conformal mapping from D_1 onto D_2 if ϕ is a 1-1 conformal mapping from D_1 onto D_2 and a homeomorphism from $\overline{D}_1$ onto $\overline{D}_2$.

We need the following fundamental theorem (see e.g., Curtiss [3] Theorem 13.7.1).

<u>Theorem A</u>. (Extended Riemann Mapping Theorem). Let Γ be a Jordan curve. Then there exist extended conformal mappings Ψ and Ψ^* from $Int(\Gamma)$ onto B and from $Ext(\Gamma)$ onto B^*, respectively.

Since D is a Jordan domain, we can write

$$\partial D = \{\Gamma_0, \Gamma_1, \cdots, \Gamma_n\},$$

where $\Gamma_i (i = 0,1,\cdots,m)$ are disjoint Jordan curves and $D \subset Int(\Gamma_0)$.

<u>Lemma 1</u>. For each $k = 0,1,\cdots,m$, there exists an extended conformal mapping ϕ_k from D onto a bounded C^∞ domain D_k with $\phi_k(\Gamma_k) = S$.

<u>Proof.</u> Since $D = Int(\Gamma_0) \cap Ext(\Gamma_1) \cap \cdots \cap Ext(\Gamma_m)$, we can use Theorem A $(m + 1)$ times to get the desirable extended conformal mapping ϕ_k (put k as the last time and notice that the conformal mapping maps an interior C^∞ curve into a C^∞ curve). We leave the details to readers. ∎

It is easy to see by definition that for any extended conformal mapping ϕ from a bounded domain D onto E, there exists a constant $C > 0$ such that for all $x, y \in \overline{D}$,

$$\frac{1}{C}|x-y| \le |\phi(x)-\phi(y)| \le C|x-y|. \tag{3}$$

Thus we can take a constant $C = C(D) > 0$ such that (3) holds for each ϕ_k, $k = 0,1,\cdots,m$, especially, we have for all $k = 0,1,\cdots,m$, $x \in D$,

$$\frac{1}{C}\rho(x) \le \rho_k(\phi_k(x)) \le C\rho(x), \tag{4}$$

where $\rho_k(y) = dist\ (y, \partial D_k)$.

<u>Lemma 2</u>. Let ϕ be an extended conformal mapping from the Jordan domain D onto E. Then we have

$$G_D(x,y) = G_E(\phi(x),\phi(y)), \quad x, y \in D.$$

<u>Proof</u>. Since for a Jordan domain D, $G_D(\cdot,\cdot)$ is characterized by the following conditions: (due to the uniqueness of the Dirichlet problem) For any $y \in D$, we have

(i) $G_D(\cdot,y)$ is positive and harmonic in $D\backslash\{y\}$;

(ii) $\lim_{x\to\partial D} G_D(x,y) = 0$;

(iii) $\frac{1}{\pi}\ell n\frac{1}{|x-y|} - G_D(x,y)$ is bounded for $x \in D$.

So we need only verify them to the function $f(x,y) \equiv G_E(\phi(x),\phi(y))$ for D.

Since a harmonic function remains harmonic after any analytic change of variable, (see e.g., [4] Theorem 7.3) (i) holds for $f(x,y)$. (ii) follows by the definition. For (iii), we have

$$\frac{1}{\pi}\ell n\frac{1}{|x-y|} - f(x,y)$$

$$= \frac{1}{\pi}\ell n\frac{|\phi(x)-\phi(y)|}{|x-y|} + \left[\frac{1}{\pi}\ell n\frac{1}{|\phi(x)-\phi(y)|} - G_E(\phi(x,)\phi(y))\right]. \tag{5}$$

Since $G_E(\cdot,\cdot)$ satisfies (iii) for E, the second term of (5) is bounded for $x \in D$. The first term of (5) is bounded by (3). ∎

<u>Lemma 3</u>. For all $x,y \in B$, we have

$$\frac{1}{2\pi}\ell n\left[1 + \frac{\rho_B(x)\rho_B(y)}{|x-y|^2}\right] \le G_B(t,y) \le \frac{1}{2\pi}\ell n\left[1 + 4\frac{\rho_B(x)\rho_B(y)}{|x-y|^2}\right], \tag{6}$$

and for $R > 1$, $x,y \in B^* \cap B(0,R)$, we have

$$\frac{1}{2\pi}\ell n\left[1 + \frac{\rho_{B*}(x)\rho_{B*}(y)}{|x-y|^2}\right] \le G_{B*}(x,y)$$

$$\le \frac{1}{2\pi}\ell n\left[1 + 4R^2\frac{\rho_{B*}(x)\rho_{B*}(y)}{|x-y|^2}\right]. \tag{7}$$

<u>Proof</u>. By the formula of the Green function of B, we have

$$G_B(x,y) = \frac{1}{\pi}\ell n\frac{1}{|x-y|} - \frac{1}{\pi}\ell n\frac{1}{m|x-y|} = \frac{1}{2\pi}\ell n\left(\frac{m(x,y)^2}{|x-y|^2}\right), \tag{8}$$

where $m(x,y) = |y|\ |x - \frac{y}{|y|^2}|$.

By a simple calculation (see Chung [1] or Zhao [6]) we have

$$m(x,y)^2 = |x-y|^2 + (1-|x|^2)(1-|y|^2)$$
$$= |x-y|^2 + \rho_B(x)\rho_B(y)(1+|x|)(1+|y|). \tag{9}$$

Thus (6) follows from (8) and (9).

For $x,y \in B^* \cap B(0,R)(R>1)$ we have $x^* \equiv \frac{x}{|x|^2} \in B$ and

$$\rho_B(x^*) = 1 - \frac{1}{|x|} = \frac{|x|-1}{|x|} = \frac{\rho_{B*}(x)}{|x|},$$

$$|x^*-y^*| = \frac{|x-y|}{|x|\,|y|}.$$

Hence,

$$\frac{\rho_B(x^*)\rho_B(y^*)}{|x^*-y^*|^2} = \frac{\rho_B{}^*(x)\ \rho_{B*}(y)|x|\,|y|}{|x-y|^2}. \tag{10}$$

The inequalities (7) follows from the equality $G_{B*}(x,y) = G_B(x^*,y^*)$, (6) and (10). ■

<u>Lemma 4</u>. (i) $\ell n(1+a) \le a$ for $a \in (0,\infty)$;

(ii) $\frac{a}{2} \le \ell n(1+a)$, for $a \in (0,\frac{1}{2}]$;

(iii) for any $\beta>0$ there exists a constant $C = C(\beta)>0$ such that for all $a \in (0,\infty)$, $\frac{1}{C}\ell n(1+a) \le \ell n(1+\beta a) \le C\ell n(1+a)$.

<u>Proof</u>. By elementary claculus. ■

<u>Lemma 5</u>. If D is a bounded $C^{1,1}$ domain in R^2, then for any $\delta > 0$, there exists $C = C(D,\delta) > 0$ such that for all $x,y \in D$ with $|x-y| \ge \delta$, we have

$$G_D(x,y) \le C\rho(x)\rho(y). \tag{11}$$

Proof. Take $\gamma>0$ such that $D \subset B(0,\gamma)$. By a simple calculus or see Chung [ ] (28) and (32), we have

$$G_D(x,y) \le G_{B(0,\gamma)}(x,y) \le \frac{1}{\pi}\ell n\frac{3\gamma}{|x-y|}. \tag{12}$$

By using the same method as Widman [ ] did for $d \ge 3$, (only replace the inequality $G_D(x,y) \le \frac{C_d}{|x-y|^{d-2}}$ there by (12)) we have that there is some $A > 0$,

$$G_D(x,y) \le \frac{A}{\pi}(\ell n\frac{3\gamma}{|x-y|})\frac{\rho(x)\rho(y)}{|x-y|^2}.$$

Then (11) follows from the above inequality. ■

Lemma 6. If D is a bounded $C^{1,1}$ domain in R^2, then for any two sequences $\{x_n\}$, $\{y_n\} \subset D$ such that $x_n \to x_0 \in \overline{D}$ and $y_n \to y_0 \in \overline{D}$ with $x_0 \ne y_0$, there exists the limit:

$$0 < \lim_n \frac{G(x_n,y_n)}{\rho(x_n)\rho(y_n)} < \infty. \tag{13}$$

Proof. We obtain (13) by the same arguments as those in Zhao [7] Lemma 1 (the statement there is for $d \ge 3$). We need only replace Widman inequality $G(x,y) \le C\frac{\rho(x)\rho(y)}{|x-y|^d}$ $(d \ge 3)$ by (10), since the Poisson formula for a ball used there has the same form for $d \ge 2$. ■

Proof of Theorem 1. If the assertion is not true, then there must be two sequences $\{x_n\}$ and $\{y_n\} \in D$ such that

$$I(x_n,y_n) \equiv \frac{G_D(x_n,y_n)}{\ell n\left(1 + \frac{\rho(x_n)\rho(y_n)}{|x_n-y_n|^2}\right)} \longrightarrow \infty \text{ or } \longrightarrow 0. \tag{14}$$

We can assume that $x_n \to x_0 \in \overline{D}$ and $y_n \to y_0 \in \overline{D}$.

Case (i) $x_0 \neq y_0$ and $x_0 \in D$, $Y_0 \in D$.

In this case we have

$$I(x_n, y_n) \to I(x_0, y_0)$$

and $I(x_0, y_0)$ is a strictly positive finite number. This is a contradiction to (14).

Case (ii) $x_0 \neq y_0$ and x_0(or y_0) $\in \partial D$.

Since $\dfrac{\rho(x_n)\rho(y_n)}{|x_n - y_n|^2} \longrightarrow 0$, we have

$$\frac{\rho(x_n)\rho(y_n)}{\ell n\left(1 + \dfrac{\rho(x_n)\rho(y_n)}{|x_n - y_n|^2}\right)} \longrightarrow |x_0 - y_0|^2. \tag{15}$$

By Lemma 2 and Lemma 6, we have

$$\ell im_n \frac{G_D(x_n, y_n)}{\rho_0(\phi_0(x_n))\rho_0(\phi_0(y_n))} = \ell im_n \frac{G_{D_0}(\phi_0(x_n), \phi_0(y_n))}{\rho_0(\phi_0(x_n))\rho_0(\phi_0(y_n))} > 0. \tag{16}$$

We now lead to a contradiction to (14) by (15), (4) and (16).

Case (iii) $x_0 = y_0 \in D$.

Since $\dfrac{\rho(x_n)\rho(y_n)}{|x_n - y_n|^2} \longrightarrow \infty$, we have

$$\frac{\frac{1}{\pi}\ell n\dfrac{1}{|x_n - y_n|}}{\ell n\left(1 + \dfrac{\rho(x_n)\rho(y_n)}{|x_n - y_n|^2}\right)} \longrightarrow \frac{1}{2\pi}. \tag{17}$$

By the properties of the Green function G_D, we have that in this case,

$$\frac{G_D(x_n, y_n)}{\frac{1}{\pi}\ell n\dfrac{1}{|x_n - y_n|}} \longrightarrow 1. \tag{18}$$

Thus we get a contradiction to (14) by (17) and (18).

Case (iv) $x_0 = y_0 \in \partial D$.

Suppose $x_0 = y_0 \in \Gamma_k$, $k = 0,1,2,\cdots,m$. Put

$$U = \begin{cases} B, & \text{if } k = 0 \\ B^*, & \text{if } k = 1,\cdots,m \end{cases}$$

and take $R > 1$ such tat $\overline{D}_k \subset B(0,R)$, $k = 0, 1,\cdots,m$. Then we have $D_k = \phi_k(D) \subset U$ and $\partial U = S = \phi(\Gamma_k)$. For the C^∞ domain D_k, we have by the Green identity, the Poisson kernel of D_k:

$$K_{D_k}(v,z) = \frac{1}{2}\frac{\partial}{\partial n_z}G_{D_k}(v,z),$$

$v \in D_k$, $z \in \partial D_k$. Hence it is easy to see by the properties of the Green function that for u, $v \in D_k$,

$$G_{D_k}(u,v) = G_u(u,v) - \frac{1}{2}\int_{\partial D_k|\partial U} G_U(u,z)\frac{\partial}{\partial n_z}G_{D_k}(u,z)\sigma(dz)$$

$$= G_U(u,v) - \frac{1}{2}H(u,v). \tag{19}$$

Put $\delta = \text{dist}(\partial U, \partial D_k\backslash\partial U)$. For any u, $v \in D_k$ with $\text{dist}(u, \partial U) < \frac{\delta}{2}$ and $\text{dist}(u,\partial U) < \frac{\delta}{2}$, by Lemma 3, there is some $C_1(\delta) > 0$ such that

$$G_U(u,z) \leq C_1(\delta)\rho_k(u) \tag{20}$$

and by Lemma 5, there is some $C_2(\delta) > 0$ such that ($C_1(\delta)$ and $C_2(\delta)$ may depend on D_k, i.e., on D)

$$\frac{\partial}{\partial n_z}G_{D_k}(v,z) \leq C_2(\delta)\rho_k(v). \tag{21}$$

Set $u_n = \phi_k(x_n)$, $v_n = \phi_k(y_n)$. Then we have

$$u_n, v_n \rightarrow \phi_k(x_0) = \phi_k(y_0) \in \phi_k(\Gamma_k) = \partial U.$$

By (20) and (21), we have

$$\frac{H(u_n, v_n)}{\ell n\left(1 + \frac{\rho_k(u_n)\rho_k(v_n)}{|u_n - v_n|^2}\right)} \longrightarrow 0. \tag{22}$$

By (19), (22), Lemma 3 and Lemma 4 (iii), there exists a constant $C_1 > 0$ and $N \geq 1$ such that for all $n \geq N$.

$$\frac{1}{C_1} \leq \frac{G_{D_k}(u_n, v_n)}{\ell n\left(1 + \frac{\rho_k(u_n)\rho_k(v_n)}{|u_n - v_n|^2}\right)} \leq C_1. \tag{23}$$

By (3), (4) and Lemma 4 (iii), there exists $C_2 > 0$ such that for all $n \geq 1$,

$$\frac{1}{C_2} \leq \frac{\ell n\left(1 + \frac{\rho_k(u_n)\rho_k(v_n)}{|u_n - v_n|^2}\right)}{\ell n\left(1 + \frac{\rho(x_n)\rho(y_n)}{|x_n - y_n|^2}\right)} \leq C_2. \tag{24}$$

Since $G_D(x_n, y_n) = G_{D_k}(u_n, v_n)$ by Lemma 2, it follows from (23) and (24) that for all $n \geq N$,

$$\frac{1}{C_1 C_2} \leq \frac{G_D(x_n, y_n)}{\ell n\left(1 + \frac{\rho(x_n)\rho(y_n)}{|x_n - y_n|^2}\right)} \leq C_1 C_2.$$

This is a contradiction to (14). ∎

Proof of Theorem 2. Due to Theorem 1, we need only prove the following inequality:

$$Q(x,y,z) \equiv \frac{\ell n\left(1 + \frac{\rho(x)\rho(y)}{|x-y|^2}\right) \ell n\left(1 + \frac{\rho(y)\rho(z)}{|y-z|^2}\right)}{\ell n\left(1 + \frac{\rho(x)\rho(z)}{|x-z|^2}\right)}$$

$$\leq C[F(x,y) + F(y,z)]. \tag{25}$$

Recall here $F(x,y) = \max[\ell n\frac{1}{|x-y|}, 1]$.

Since $\rho(x)$ is bounded in D, we have for all $x,y \in D$,

$$\ell n\left(1 + \frac{\rho(x)\rho(y)}{|x-y|^2}\right) \leq CF(x,y). \tag{26}$$

By symmetry, we can assume that

$$|x-y| \leq |y-z|. \tag{27}$$

Hence we have

$$|x-z| \leq |x-y| + |y-z| \leq 2|y-z|. \tag{28}$$

We also need the following simple inequality: for all $x,y \in D$,

$$|\rho(x)-\rho(y)| \leq |x-y|. \tag{29}$$

We divide the whole situation into 3 cases:

Case (i) $\frac{\rho(x)\rho(y)}{|x-y|^2} \geq \frac{1}{2}$.

By (29), we have

$$\frac{1}{2} \leq \frac{\rho(x)[\rho(x)+|x-y|]}{|x-y|^2} = \left[\frac{\rho(x)}{|x-y|}\right]^2 + \frac{\rho(x)}{|x-y|}.$$

So we have

$$\frac{1}{3} \leq \frac{\rho(x)}{|x-y|}. \tag{30}$$

Hence by (28) and (30),

$$\frac{\rho(y)\rho(z)}{|y-z|^2} \leq \frac{4[\rho(x)+|x-y|]\rho(z)}{|x-z|^2} \leq 16\frac{\rho(x)\rho(z)}{|x-z|^2}. \tag{31}$$

Using (31), Lemma 4 (iii) and (26), we obtain

$$Q(x,y,z) \le C\ell n\left(1 + \frac{\rho(x)\rho(y)}{|x-y|^2}\right) \le CF(x,y).$$

Thus (25) holds in this case.

Case (ii) $\dfrac{\rho(x)\rho(y)}{|x-y|^2} < \dfrac{1}{2}$ and $\dfrac{\rho(x)\rho(z)}{|x-z|^2} \ge \dfrac{1}{2}$.

By (26),

$$Q(x,y,z) \le \frac{\ell n\frac{3}{2}}{\ell n\frac{3}{2}}\ell n\left(1 + \frac{\rho(y)\rho(z)}{|y-z|^2}\right) \le CF(y,z).$$

Then (25) holds.

Case (iii) $\dfrac{\rho(x)\rho(y)}{|x-y|^2} < \dfrac{1}{2}$ and $\dfrac{\rho(x)\rho(z)}{|x-z|^2} < \dfrac{1}{2}$.

We first prove that

$$\frac{\rho(y)}{|x-y|} \le 2. \tag{32}$$

If not, by the first condition, we have

$$\frac{\rho(x)}{|x-y|} \le \frac{1}{4}.$$

Then by (29),

$$\rho(y) \le \rho(x) + |x-y| \le \frac{5}{4}|x-y|,$$

which is a contradiction. Hence (32) is true in this case.

By the second condition and Lemma 4 (i) (ii), we have

$$Q(x,y,t) \le \frac{\dfrac{\rho(x)\rho(y)}{|x-y|^2} \cdot \dfrac{\rho(y)\rho(z)}{|y-z|^2}}{\dfrac{1}{2} \qquad \dfrac{\rho(x)\rho(z)}{|x-z|^2}}$$

$$= 2\frac{\rho(y)^2 \quad |x-z|^2}{|x-y|^2 \quad |y-z|^2}$$

$$\le 2\times 4\times 4 = 32.$$

The last inequality is due to (28) and (32). Then (25) holds in the last case. ∎

References

[1] K. L. Chung, Green's function for a ball, Seminar on Stochastic Processes 1986, 1-13. Birkhauser, Boston, 1987.

[2] M. Cranston, E. Fabes, and Z. Zhao, Potential theory for the Schrödinger equation, to appear in Transactions of the American Mathematical Society.

[3] J. H. Curtiss, Introduction to Functions of a Complex Variable, New York and Basel, 1978.

[4] F. P. Greenleaf, Introduction to Complex Variables, Philadelphia, 1972.

[5] K-O. Widman, Inequalities for Green functions of second order elliptic operators, Report No. 8, 1972, Department of Mathematics, Linkoping University, 1972.

[6] Z. Zhao, Conditional gauge with unbounded potential, Z. Wahrsch. Verw. Gebiete 65, (1983), 13-18.

[7] Z. Zhao, Green function for Schrödinger operator and conditioned Feynman-Kac gauge, Journal of Mathematical Analysis and Applications, Vol. 116, No. 2, 1986, 309-334.

Zhongxin Zhao
Institute of System Science
Academia Sinica
Beijing, China

CORRECTION

by

Z.R. Pop-Stojanović

In my paper [1] please enter the following correction:

page 191, lines -9,-8: Let z ... . Then...y → z.

page 192, line + 4: Erase the first sentence of the Theorem. Start the statement of the Theorem with "The... in the second sentence.

[1] Z.R. Pop-Stojanović, Last Exit time and Harmonic measure for Brownian motion in R^d , Seminar on Stochastic Processes 1986, 191-194, Birkhauser, Boston 1987.

Z.R.Pop-Stojanović
Department of Mathematics
University of Florida
Gainesville, Florida 32611

Progress in Probability and Statistics

1 ÇINLAR/CHUNG/GETOOR. Seminar on Stochastic Processes, 1981
ISBN 3-7643-3072-4
2 KESTEN. Percolation Theory for Mathematicians
ISBN 3-7643-3107-0
3 ASMUSSEN/HERING. Branching Processes
ISBN 3-7643-3122-4
4 CHUNG/WILLIAMS. Introduction to Stochastic Integration
ISBN 3-7643-3117-8
5 ÇINLAR/CHUNG/GETOOR. Seminar on Stochastic Processes, 1982
ISBN 3-7643-3131-3
6 BLOOMFIELD/STEIGER. Least Absolute Deviation
ISBN 3-7643-3157-7
7 ÇINLAR/CHUNG/GETOOR. Seminar on Stochastic Processes, 1983
ISBN 3-7643-3293-X
8 BOUGEROL/LACROIX. Products of Random Matrices with Application to Schrodinger Operator
ISBN 3-7643-3324-3
9 ÇINLAR/CHUNG/GETOOR. Seminar on Stochastic Processes, 1984
ISBN 3-7643-3327-8
10 KIFER. Ergodic Theory of Random Transformations
ISBN 3-7643-3319-7
11 EBERLEIN/TAQQU. Dependence in Probability and Statistics
ISBN 3-7643-3323-5
12 ÇINLAR/CHUNG/GETOOR. Seminar on Stochastic Processes, 1985
ISBN 3-7643-3331-6
13 ÇINLAR/CHUNG/GETOOR/GLOVER. Seminar on Stochastic Processes, 1986
ISBN 3-7643-3353-7
14 DEVROYE. A Course in Density Estimation
ISBN 3-7643-3365-0
15 ÇINLAR/CHUNG/GETOOR/GLOVER. Seminar on Stochastic Processes
ISBN 3-7643-3381-2
16 KIFER. Random Perturbations of Dynamical Systems
ISBN 3-7643-3384-7